Pra

‘Both inspirational and richly human, the book offers a compelling picture of science at the cutting edge’

Times Higher Education Supplement

‘Anna Buckley, the producer of *The Life Scientific*, distils the scientists’ passions skilfully into print’

Sunday Times, Science Books of the Year 2018

Praise for *The Life Scientific*

‘*The Life Scientific* does that clever thing, making a difficult subject more accessible not by talking down to us but by talking to scientists who are so passionate about their subject that they entice us in’

Kate Chisholm, *Spectator*

‘Each week Jim Al-Khalili interviews brilliant scientists about their brilliant lives in research. It’s inspiring because their achievements are huge’ *The Times*

‘*The Life Scientific* . . . transmutes the explanation of ideas into discovery. The listener always feels in the same room as the speakers. In showing non-scientists why science offers so many paths to discovery it has no equal’

Gillian Reynolds, *Telegraph*

‘This excellent series specialises in guests who are both fascinating and admirable’

David Crawford, *Radio Times*

ANNA BUCKLEY is the series producer of *The Life Scientific* on BBC Radio 4. She has worked with the presenter Jim Al-Khalili since the programme was launched in 2011 and has produced nearly a hundred interviews with leading scientists, revealing the men and women behind the latest scientific discoveries. She has worked in the BBC Radio Science Unit for twenty years and lives in London with her husband, environmental consultant Mike Quint and two teenage daughters.

PROFESSOR JIM AL-KHALILI OBE FRS is a theoretical physicist, author and broadcaster based at the University of Surrey. He has written eleven books and has presented *The Life Scientific* since 2011. He is a recipient of the Royal Society Faraday Prize, the Institute of Physics Kelvin medal and the inaugural Stephen Hawking Medal for Science Communication.

THE LIFE SCIENTIFIC
EXPLORERS

ANNA BUCKLEY

WITH A FOREWORD BY JIM AL-KHALILI

First published in Great Britain in 2018 by Weidenfeld & Nicolson
This paperback edition published in 2020 by Weidenfeld & Nicolson
an imprint of The Orion Publishing Group Ltd
Carmelite House, 50 Victoria Embankment
London EC4Y 0DZ

An Hachette UK Company

1 3 5 7 9 10 8 6 4 2

By arrangement with the BBC.

A CIP catalogue record for this book is available from the British Library.

ISBN (Mass Market Paperback) 978 1 4746 0809 1
ISBN (eBook) 978 1 4746 0749 0

Typeset by Input Data Services Ltd, Somerset

Printed and bound in Great Britain by Clays Ltd, Elcograf S.p.A.

FSC www.fsc.org MIX Paper from responsible sources FSC® C104740

www.orionbooks.co.uk
www.weidenfeldandnicolson.co.uk

To Mike, with love

CONTENTS

FOREWORD

At the time of writing, I have been presenting *The Life Scientific* on BBC Radio 4 for seven fantastic years. I have interviewed Nobel Prize winners, presidents of the Royal Society and Government Chief Scientific Advisors, I have chatted to engineers and inventors, probed the lives and work of astronomers and astronauts, philosophers and psychologists, entomologists, ornithologists and virologists, fungal ecologists, cosmologists, immunologists, primatologists and climatologists. Oh, the list goes on, but I fear you might think I'm just inventing 'ologies' to impress. Basically, if I could remember all the science I have picked up along the way during these seven years, I would be the ultimate science polymath. Sadly, the human brain is only finite (I know that from interviewing neurophysiologists), and anyway I don't have the world's greatest memory. I remember snippets here and there – enough to impress at dinner parties, but that's about all.

Maybe having the scientific lives of my *Life Scientific* guests in a book will help me as a handy reference manual. One thing is for sure, such a book is long overdue. But before I tell you about the author, let me say something about how the programme itself came about.

The Life Scientific was the brainchild of the then new controller of BBC Radio 4, Gwyneth Williams. Gwyneth had contacted me to ask if I might be interested in hosting a brand-new regular weekly science programme that she was keen to commission. The idea, roughly, was that I would chat to other scientists about their life and work for half an hour during that 'golden slot' at 9 a.m., following on from the hugely popular daily news show *The Today Programme*. She pointed out that Tuesday mornings were the only weekday slot without a recognisable long-running weekly programme in the schedule. I was of course very keen. I am sure the hugely experienced production team in the BBC Radio Science Unit were more than a little nervous to have imposed on them a new showcase programme from their new boss, which would be presented by an academic scientist with little radio broadcasting experience. But they knew they had to make it work.

And so the adventure began. From the start, I worked with a number of fantastic producers, and it was a steep learning curve. I remember once being told while in the studio through my headphones to repeat a question, but with a smile. I was puzzled. This was radio; what was the point of smiling. It turns out you can hear a smile, because on radio, everything is in the voice. But one producer in particular has worked closely with me, not only making the programmes but experimenting with the format, style and structure. She is the series producer and also the author of this series of books. Anna Buckley is an extraordinary talent, not only as programme maker but as a synthesiser of information and storyteller. Unlike me, with my unreliable memory, Anna could very well recite every guest we've had on *The Life Scientific* – all 170 of them – and tell you what their scientific discipline is and their claim to fame. It might be my voice you hear on the radio, but it will very likely be Anna who has edited over an hour of chat between two scientists down to a tight 28 minutes of radio gold. It's exciting to see more of her alchemy at work in this excellent book.

Jim Al-Khalili
July 2018

INTRODUCTION

Science is full of surprises, and it's the unexpected twists and turns in these intellectual life stories that I like so much. Serendipity has a starring role. Cue the appearance of a fossilised pine needle that couldn't be ignored.

Who knows what makes us fall in love? Monica Grady was seduced by the minerals in Moon rock. 'The colours were so deep and so clear.' Lucie Green was captivated by images of the sun: 'I do also love the fact that it is a contained nuclear bomb.' When Richard Fortey split open a rock on a beach in Wales as a boy and 'a perfectly formed creature popped out as if it were made for [him]', a lifelong love affair with trilobites began.

The scientists in this volume are all explorers. Inventors, detectives, activists and radicals will follow, although, of course, most scientists are a mix of all these things. Many are at the forefront of discovery, collecting the evidence on which science is based, exploring new terrain. Driven by a desire to conquer the unknown, they are opening up new worlds: from a distant moon of Saturn to the molecules in our DNA. Henry Marsh slices through the 'slightly firm white jelly' that sits between our ears.

They all travel hopefully in their minds. The journey is the reward and there is no talk of giving up. Botanist Sandy Knapp revels in the never-ending nature of her work: 'The more I find out, the more I realise I don't know, so it's a constant voyage of discovery.' Lichenologist Pat Wolseley agrees. She might have been an artist, but it was the questions that drew her back to science. 'And the questions, they never stop.'

The sheer scale of it all can be overwhelming. Distressed to discover that computers couldn't help him to sort a million galaxies according to their shape, Chris Lintott came up with an ingenious solution. He set up a website and asked as many people as possible to do his work for him. Thousands of curious people helped Chris with his PhD. And the idea of citizen science took root in the UK.

Finding things out is fun. It can also be addictive. John Sulston (who sadly died in March 2018) describes the joy of watching cells divide, eight hours a day, non-stop for a year and a half. There was a small hollow in the wooden floor created by the wheels of his chair as he repeatedly moved it back and forth. 'You have to be obsessive to do certain kinds of science,' he said. 'I'm not alone in that.' It took him seven years to sequence the genome of a tiny nematode worm. From there he graduated to human DNA. We have John to thank for Britain's involvement in the Human Genome Project, a feat of biological exploration that has been compared to the Apollo missions to the Moon. A list of three billion genetic letters as they appear on human DNA has been written out in full.

Jocelyn Bell Burnell became obsessed by a tiny smudge in a graph. Determined to understand its cause, she discovered a new kind of star. More than 2,000 pulsating radio stars have been identified since she published her results in 1968. Michele Dougherty persuaded mission control to take a closer look at the magnetic field around one of Saturn's moons when she noticed a strange wiggle in the data that couldn't be explained. Enceladus is now a prime destination in the search for extraterrestrial life.

Several of the scientists featured in this book are involved in the search for life on Mars: 'Primitive life is possible, but it will be very small and difficult to find.' Monica Grady learnt her trade (analysing the chemical composition of meteorites) from the late Colin Pillinger. Colin's *Beagle* 2 lander, destined to land on Mars on Christmas Day 2003, enthused the nation. Damien Hirst and the Brit pop band Blur got involved. As did the Queen. 'Every generation needs its gothic cathedrals,' he said. 'We shouldn't be doing things that everyone else has done already.' When the robot *Philae* landed on a speeding comet four billion miles from home in November 2014, it was broadcast around the world, and the chief scientist involved, Matt Taylor, became a media star.

The space scientists were wonderfully down to earth. Brian Cox needs to have his hand held when he crosses the road, according to comedian Robin Ince. He supports the idea that there are many universes, not just one. Quantum mechanics is the way to visit them. The mathematicians Marcus du Sautoy and Eugenia Cheng travel further still, sitting at home or in hotel bars imagining abstract worlds. Why limit yourself to three dimensions (or four if you include time) when you could be enjoying infinitely more?

Some have travelled to the ends of the Earth. Jane Francis loves camping on Antarctic ice. Richard Fortey slipped into the Arctic ocean, weighed down by a rucksack full of rocks. Hazel Rymer measures microgravity on volcanoes. Helen Sharman was the first British person to go into space. Others are more sedate. Having enjoyed plenty of far-flung geological adventures as a young man, Sanjeev

Gupta turned his attention to Mars, imagining it could be studied 'in the kitchen over a glass of wine'. He decides where the Martian *Curiosity* rover should go next.

Others travel back in time, recreating vanished worlds. What did our planet look like hundreds of millions of years ago? How were the ancient continents arranged before they came together to form the latest supercontinent Pangaea? Nick Fraser has unearthed some crazy-looking reptiles from the Triassic that make the wacky creatures drawn by Dr Seuss look tame.

I would like you to treat this book like a pick and mix. There is no need to read everything at once. If you develop a taste for science, then my job is done. If you discover something new, I will be delighted. But please don't feel obliged to laboriously understand everything. The point is not to know it all. If Nick Fraser can happily admit that he felt 'out of his depth going back another hundred million years', then so can you. A distressing number of people seem to worry that science is beyond their understanding, but no one understands everything. And, as Eugenia Cheng reminds us: 'You can enjoy an art gallery, without being able to paint.' No prior knowledge is required to read this book.

Working as the series producer of *The Life Scientific*, I have no hierarchy in mind when I sit down on a Friday morning, or wake up in the middle of the night, wondering who Jim should interview next. Science is a collective endeavour conducted over centuries. Anyone who is curious can make a contribution. Who am I to judge one person to be 'more important' than another? Instead, we aim to offer a smorgasbord of scientific ideas and a range of scientists, at different stages in their careers. Their ambitions vary, as does the degree of recognition they have received. Those featured on the programme have been chosen, first and foremost, for the stories they tell.

I have been lucky enough to talk at length to almost all the scientists in this volume, when preparing myself, and them, for their interview with Jim. On more than one occasion, I'm sorry to say, I have phoned a future guest on the programme at the time agreed, exhausted by the stresses of the week and ready to fall asleep. But after talking for an hour, sometimes more, I have often been transformed, energised, exhilarated and inspired by a scientist who has pursued his or her passion with exuberance or quiet zest. I hope that one, or more, of the stories collected in this volume might have a similar effect on you.

AUTHOR'S NOTE

The chapters that follow are based on interviews that were broadcast on *The Life Scientific* between September 2011 and January 2018 on BBC Radio 4, many of which I produced. They aim to capture what was said and something of the spirit in which it was said. They are not comprehensive profiles. Few people's lives fit neatly into 27 minutes and 40 seconds, and I am always struck by just how little can be said in this amount of time. (After seven years of producing *The Life Scientific*, you'd think I might have learnt by now.)

In the earlier episodes we recorded interviews with people who knew the guest or their work and played a clip into the interview for the guest to comment on, so sometimes other voices will pop up. I have also added details and tried to provide necessary context. When preparing for each new interview, I spend days researching the life and work of the guest and hours talking to them on the phone, working out what the story of their life scientific might be. Some of the information gleaned in this way has been included. If you enjoyed listening to these programmes when they were on air, I hope you will find some added value here.

Science moves on, sometimes rapidly, and I am painfully aware of everything that has happened since these interviews were recorded. I forced myself to stay true to the original conversation, but occasionally the opportunity to update the story was irresistible.

MONICA GRADY

'I am interested in the stories meteorites bring with them'

Grew up in: Leeds
Home life: married to a fellow space scientist; they have a son called Jack and a grandson, Matthew
Occupation: space scientist
Job title: Professor of Planetary and Space Sciences, the Open University, Milton Keynes
Inspiration: seeing Moon rock through a microscope
Passion: extraterrestrial rock
Mission: to open everyone's eyes to what's out there
Best moment: 'There are too many'
Worst moment: getting fired within an hour of being hired for breaking the mass spectrometer
Advice to young scientists: 'Do what you enjoy'
Date of broadcast: 16 October 2012

Struck by the beautifully coloured minerals in a piece of Moon rock that had been brought down to Earth by Neil Armstrong, Monica Grady became a planetary geologist, specialising in meteorites. Every rock tells a story, and stories from space feel particularly special. She runs the global database of all the extraterrestrial rocks that have been found on Earth and has studied two Martian meteorites in great detail, EETA 79001 and Allan Hills 84001. In a landmark paper in 1989, she announced that organic molecules had been found in EETA 79001, and hopes were raised that there might once have been life on Mars. Microbial, not little green men.

Monica Grady is the queen of alien rock arrivals.[1] 'I've been known as Monica Grady Meteorite Lady for quite some time now,' she said, clearly happy with the description.

For many years, Monica was the curator of the largest meteorite collection in the world at the Natural History Museum in London, which is home to 5,000 specimens. An artist who visited Monica at the museum described her as 'a bit like a meteorite herself. She was just so full of energy. She seemed to know all their stories, which was mind-bending. They were like her children.'

'I don't know about that,' Monica said, laughing, 'but they're certainly more obedient than my son.' The rocks 'sit there and do what they're told'.

As a child, Monica enjoyed exploring the Yorkshire Dales. 'It doesn't take you long to find a limestone pavement,' she said. Walking on the wobbly blocks of weathered rock at Malham and avoiding the hazardous grykes between them, Monica, aged eight, wanted to know what had made this other-worldly landscape. How had it come to be the way it was?

'Were you bringing rocks home, even then?' Jim asked.

'I was bringing home lots of bits and pieces, yes. All sorts of things – the odd fossil that I found, bits of shell.

1 About 100,000kg of space rock is deposited on Earth every day. Most of it arrives as dust and gravel, but potato-sized meteorites are not uncommon. A meteorite the size of a double bed stands proud in a dusty Namibian field.

For ages and ages and ages, I had a bit of sea-washed glass that I'd found on the beach.' She imagined this pretty stone was a precious gem and only discovered much later that it must have started life as piece of broken glass.

In the final year of her joint honours degree in chemistry and geology at Durham University, a lecturer arranged for the students to see a slice of the Moon, courtesy of a teaching pack donated by the *Apollo 11* mission. Thinly sliced, polished and placed under a strong polarising microscope, Moon rock looked magnificent. 'The colours were so deep and so clear, beautiful cerise and turquoise, and pink and yellow and beige and grey,' Monica said, her voice becoming soft and dreamy. The rock was a patchwork of sophisticated translucent colours, shining brightly. 'Some of them were a bit cracked but you can see such sharp outlines of the crystals.' There were clusters of crystals with sharply defined edges, like limbal rings around an iris.

The rock she was looking at had come from magma, 'from a lava that's crystallised'. And she could 'see how all the grains fit together so beautifully'. It was stunning aesthetically and a scientific coup – a close-up of the Moon. 'Those thin sections of the *Apollo* rocks were what really inspired me,' she said.

'Do you ever find yourself looking at the ground, hoping to find meteorites?' Jim asked.

'Oh yes,' Monica exclaimed. 'All the time!'

There could be meteorites anywhere on Earth and Monica has joined many meteorite-hunting expeditions. Antarctica is a particularly good place to look. Blackened meteorites on white ice are hard to miss, and the arid conditions mean the meteorites are well preserved.

'Can you tell, just by looking at it, where a meteorite has come from?'

'I've got a very small chip of one here,' Monica said, whipping a small rock out of her pocket. 'See if you can look at it and tell me where you think that one came from?'

'Can I hold it?' Jim asked, taking off his glasses. 'Well, it's a little black piece of rock,' he said. 'It's very shiny. Half of its surface is very black, which presumably is where it's burnt, coming into the atmosphere . . .'

Monica could wait no longer. 'This one's come from Mars,' she said.

'How exciting!'

'Don't drop it!'

'How much is it worth?' Jim asked. 'Can I keep it?'

'No, you can't keep it!' Monica exclaimed, reaching forward to grab it back. 'And I'm not telling you how much it's worth. It's scientifically priceless.'

Some meteorites contain 'dust from the beginning of the solar system'. Viewed from the outside, they are nothing special. The burnt remains of their journey through the Earth's atmosphere cover the surface with a blackened crust. But when they are split in two, the insides can be marvellous.

Monica encouraged Jim to look more closely at the Martian meteorite he coveted, 'a piece of a rock from a volcano on Mars which has solidified'. 'When you look closer inside,' Monica said, 'you can see minerals that have been formed by water running across the surface of Mars. You've got these secondary minerals called carbonates.'

'And this is what you do?'

'This is what I do. Yes.'

When studying the chemical composition of the minerals found in meteorites, Monica is particularly interested in carbon. 'Life is based on carbon,' she explained. 'Life is based on the DNA molecule,' she added more precisely. 'I'm not looking for DNA. I'm looking for the original bits that would go together to make a DNA molecule.' In other words, one or some of the five elements from which the DNA molecule is made: carbon, oxygen, hydrogen, nitrogen and phosphorus.

'The very first stages?' Jim said.

'The first stages, yeah. Building blocks.'

'It all sounds very sensible, but how sensible is it to even expect to find life on Mars?' Jim asked. 'After all, it is pretty inhospitable.'

'Mars is an inhospitable, cold, arid planet now,' Monica replied. 'But it wasn't always like that.' And planet Earth hasn't always been the paradise for life that it is now. 'For the first 2 billion years of Earth's history, there was nothing there. There were no major life forms.' As the Earth cooled, its atmosphere thickened and the surface of the Earth 'became really good, full of these hospitable niches where life could get going'.

It was a different story on Mars. Being just over half the size of planet Earth, the gravitational pull Mars exerted on its atmosphere was weaker, and when Mars cooled it lost its atmosphere and was no longer able to retain any surface water. And at that point in Martian history, 'the possibility of evolved complex life forms arising on Mars more or less disappeared.'

'So, going back several billion years, just as there were primitive life forms on Earth, there could well have been similar life forms on Mars?'

'When life was getting going on Earth, life could have been getting going on Mars,' Monica confirmed. 'That is the hope that keeps people looking.'

Traces of these ancient life forms have been found in

terrestrial rocks. What Monica and others want to know is: can similar traces be found in rocks on Mars? The question sounds simple enough but, as Monica said: 'It's not an easy search.'

'When life was getting going on Earth, life could have been getting going on Mars. That is the hope that keeps people looking'

When Giovanni Schiaparelli noticed a neat series of lines on the surface of Mars in 1877, many imagined aliens building canals to irrigate their crops, and the possibility of life on Mars has created waves of excitement ever since. At the beginning of the twenty-first century, our hopes of finding life on this inhospitable planet have been revitalised once again.

'Why is that?' Jim asked.

'It's because we know so much more about life on Earth.' Bacteria have been found living in black smokers at the bottom of the deepest oceans, thriving under enormous pressure in superheated hydrothermal fluids. We have discovered organisms surviving in severely acidic and dangerously alkaline conditions, on a par pH-wise with battery acid and bleach.

Just knowing Earth-bound extremophiles exist makes looking for life on a planet as inhospitable as Mars feel a little bit less flawed. Conditions that were once thought to be lethal to life have been shown to be habitable, at least for certain life forms.

'When scientists say they're looking for life on Mars,' Jim said, 'what kind of life are they talking about?'

'I'm sorry – I must say this. It's life, Jim, but not as we know it!' Monica laughed. 'I'm so, so sorry. I couldn't resist!'

In fact, scientists are looking for life 'as we know it' – carbon-based and dependent on water. 'We're looking for microbial life,' and it's not going to be easy to find. 'It will be microscopic.' It's going to be in hidden niches, beneath the ice or in shady cracks. 'It's going to be below the ground. It's going to be in caves.' It might live under sheets of ice. Anywhere that offers protection from the intense electromagnetic radiation in space. 'It's going to be hidden from ready view.'

'And presumably it's going to be trapped in rocks?'

'Yes. There will be evidence for it trapped in rocks. There might be evidence for it in ice.'

'Could there be life on Mars today?' Jim asked.

'Primitive life on Mars is possible,' Monica said. 'But it will be very small and very tricky to find.'

'So primitive microbes, not little green men?'

> *'Mars is not the kind of place where you would expect to find highly evolved life'*

'I suspect, and most people suspect, that that is the case,' Monica said. 'Mars is not the kind of place where you would expect to find highly evolved life.'

Monica spent most of the 1980s studying Martian meteorites EETA 79001 and ALH 84001.[2] She was working at the Open University at Milton Keynes with Ian Wright (a fellow geologist who liked to wear his jumpers inside out and who later became Monica's husband) and Colin Pillinger.

In 1989, this close-knit and energetic team described the existence of organic molecules in Martian meteorite EETA

2 ALH 84001 is also known as the Allan Hills meteorite, after the hills in eastern Antarctica where it was found by geologists on snowmobiles in 1984.

79001 and published a paper in the prestigious peer-reviewed journal *Nature*. Five years later, they reported a high concentration of complex organic molecules in the Allan Hills meteorite as well. These are the molecules without which life, as we know it, is not possible.

Two years later, in 1996, a team of US scientists who were also working on ALH 84001 went one step further. Writing in another respected journal, *Science*, the team (which included one scientist who had previously worked with Monica, Ian and Colin) announced that they had found evidence of primitive life on Mars. The media went mad about their claims to have found a Martian microfossil.

The Stanford University Press Team tried to put things in perspective: 'If a thousand such fossils were lined up in a row they would span the dot at the end of this sentence.' Nonetheless, evidence of a Martian microbe, a hundred times smaller than a human hair, that had been dead for more than 3 billion years, was celebrated as if we had seen little green men digging canals.

Monica was sceptical. Finding microfossils in terrestrial rocks is hard enough. 'I thought it was rubbish,' she said, langhing. But then she added, more seriously, 'each thread of evidence is interesting'.

She disagreed with the way the US team had pieced the evidence together. The leap from complex organic molecules to microfossils is enormous, in terms of biological evolution, and Monica did not find sufficient evidence to support the idea that such a leap had been made. It seemed to her to be a case of 2 + 2 = 5. And she wrote a response to the paper in *Nature* entitled 'Opening a Martian Can of Worms?'

Monica, Ian and Colin had described a concentration of complex organic compounds that was unusually high for a Martian meteorite but had stopped well short of claiming the existence of biogenic material.

'I still don't believe it,' Monica said. But it's rare for the study of meteoritics to get quite so much media attention, so it wasn't all bad. 'Without doubt it attracted attention to astrobiology,' Monica said, laughing. 'It opened the doors to funding. So, in that sense, it was wonderful!'

Jim wondered if Monica had felt upstaged by the team in the USA. 'I can't help thinking if I were in your shoes at the time, I'd be panicking ever so slightly that they'd completely trumped me and my work.'

'I didn't find the fossil in the rock because I wasn't looking using that particular technique,' Monica said. 'And to be quite frank, I don't believe in one scientist trumping another scientist. This sounds hugely idealistic, but I think as long as somebody gets to the story, it doesn't really matter who does.'

'What's the verdict now on that fossil?'

'The verdict on that fossil is, we still don't know. It hasn't been ruled out, but there's more and more work going on.'

She was being polite. There are very few scientists today who would claim that the Allan Hills meteorite contains evidence of primitive life.

Jim interviewed Monica not long after the *Curiosity* arrived on Mars, the latest rover to be sent to the red planet, following *Spirit* and *Opportunity*, which had both arrived in 2004.[3]

3 *Opportunity* was still going strong in 2018, having spent a lot of time in Perseverance Valley. *Spirit* ground to a halt in 2009.

'If we really want to understand the geology of Mars we're going to have to go there rather than wait for random bits of Mars to come to us,' Jim suggested. 'And, of course, the Mars rover *Curiosity* is busy making its way across the Martian surface as we speak, looking for clues. Are you as excited by this as I am?'

'Oh, I'm incredibly excited about it!' Monica said, with even more enthusiasm than before. 'It's a huge rover, the size of a car. The biggest thing that's ever been put on the surface of Mars.'

Like *Spirit* and *Opportunity*, *Curiosity* was built to scratch and sniff the Martian surface, take measurements and send back pictures. It could also dig a little deeper and analyse samples on location. 'It's a very, very sophisticated laboratory on wheels.'

'The latest finding is this trail of smooth little pebbles trapped in rocks,' Jim said. 'How significant is that?'

'That is very significant because we have inferred the presence of running water.'

'Didn't we know that already?' Jim asked.

'Yes. We knew in theory.' The existence of water on Mars had been announced before, but the source of that water hadn't been confirmed. If, for example, the water was created in a sudden rush when perhaps a lot of ice melted, causing a flash flood, then evidence of its existence in the past should not be interpreted as evidence of a past environment that was conducive to life. The rocks gathered by *Curiosity* in Yellowknife Bay were exciting because smooth, rounded pebbles had never been seen before. Pebbles like these suggested that water must have flowed for hundreds of thousands of years, at least. It supported the idea that there were rivers on the surface of Mars and revitalised our enthusiasm for Schiaparelli's mirage.

'We haven't really seen the evidence for long periods of standing or flowing water. And this is what the water-washed pebbles are giving us.'

'Does part of you feel, "OK, we've got the evidence for the water now. Let's stop fussing about that. Let's move on and start looking for the carbon"?' Jim asked.

'Yes, partly,' Monica admitted, 'but the problem with looking for carbon is that it's going to be very, very difficult to find without extremely sophisticated instruments. *Curiosity* is great but it hasn't got absolutely the right instruments on board. And that isn't an oversight on the part of NASA. That was a specific decision.'

NASA's argument was that you need to build up the evidence gradually. 'You can only do these things bit by bit,' Monica explained, defending NASA's steady approach.

Jim pressed her a bit. 'What if you were in charge of NASA?' he asked.

'I would give more priority to the search for carbon,' she conceded.

Having studied the Martian rocks that have fallen to Earth and found them to be rich in carbon, Monica explained that she wanted to know how representative these random meteorites might be. 'I want to know how much carbon there is on the surface of Mars, or just below the surface,' she said.

'Do you worry, at all, that it's the need to be media-friendly that drives the science that's done?'

'No, I don't believe that at all. I believe scientists do the science that they want to do. NASA, although it's funded by the American government, prioritises missions based on what the scientists want to do.' The same is true of the European Space Agency, Monica said. 'It's the scientists that drive the science and explain those stories to the general public.'

'It's just that the water scientists do seem to have an easier time than you do . . .'

'My time will come,' said Monica.

The next mission to Mars is due to launch in 2018 and the search for carbon will not be a priority. 'NASA actually don't have the specialists that we have in Europe,' Monica explained. 'The gas-analysis package on the *Beagle* 2 lander was designed and built by Europeans. NASA can't do that.'

'Are we alone?' Jim asked.

'I don't know, Jim, I don't know, but it's really good fun trying to find out.'

For the scientists involved in the search for life on Mars, proof of 'a second genesis', as they like to call it, would have near-biblical significance. 'If you've got a second genesis, you could have a third, a fourth, a millionth . . .' Monica said, excited. 'It really does start to open up the possibilities of looking for a lot of life beyond the Earth.'

And how would the rest of us react, Jim wondered, if we were to find solid evidence of life on Mars? Monica was realistic. 'There will be wildly exciting headlines that will last for a day,' she said. 'A week, maximum.' Before everyone goes back to thinking, 'Now, what am I going to have for my tea?'

There was just time for a final question from Jim: 'Do you wish you could be up there on Mars?'

'Crikey. Not half!' Monica exclaimed. 'A geologist with a hammer. Just breaking a rock open and seeing straight away. Just looking! It's just not the same, doing it second-hand with a camera on a robot.'

'A manned mission to Mars has still not been ruled out,' Jim reminded Monica. 'So you never know, if they need a geologist . . .'

'I'd volunteer like a shot.'

Courtesy of Richard Fortey

RICHARD FORTEY

'A creature that was hundreds of millions of years old popped out perfectly as if it was made for me'

Grew up in: west London
Home life: married with four children
Occupation: palaeontologist
Work address: Natural History Museum, London
Inspiration: finding a perfect fossil specimen, age 14
Passion: trilobites
Mission: to recreate vanished worlds
Best moment: presenting a paper on trilobites at a conference on plate tectonics
Worst moment: falling into the Arctic ocean weighed down by a rucksack full of rocks
Advice to young scientists: persist
Date of broadcast: 28 October 2014

Richard Fortey is enthralled by trilobites, ancient marine creatures that look a bit like woodlice. For half a century he has studied little else. He found his first specimen on a beach in Wales when he was a teenager. As a student, he joined a geological expedition to Spitsbergen, in the northernmost part of Norway, and made a remarkable discovery: hundreds of previously unknown species of trilobites were buried in limestone that was 470 million years old. He spent several years reassembling their fragile fossilised remains, studying the anatomy of different species to gain clues about how and where they lived. And, aware of some intense debates about how landmasses had moved over deep geological time, he then used his knowledge of trilobites to broaden our understanding of plate tectonics.

'Tell me about your first trilobite find,' said Jim.

'I was a schoolboy,' Richard said. 'And we were on holiday in Pembrokeshire, near St David's. I knew nothing about trilobites then. But a map on the wall of the guest house said, "Ancient fossils here. Trilobites here!"' The oldest fossils in the British Isles were buried in the rocks not far away, in Nine Wells and Porth-y-rhaw, and 'there was something about this' that Richard, aged 14, 'found extraordinarily compelling'.

The next day he went down to the beach and, while his mother sat reading or knitting, he sat in the rain smashing stones, searching for evidence of ancient sea creatures that had lived 300 to 400 million years ago. 'I spent most of that holiday, in typical British weather, sitting on a beach whacking rocks,' he said. Initially even the faintest skeletal scratches were exciting. Then he had a lucky break. 'A creature that was hundreds of millions of years old popped out perfectly, as if it was made for me,' he said, as if this find had happened yesterday. 'It was an extraordinary thrill.'

Many palaeontologists wait a lifetime for 'the golden hammer' when 'you split a piece of rock and, sitting like a kipper laid out on a plate, is the perfect specimen.' More usually, scattered bits of broken bones need to be pulled together before they begin to resemble a creature. This specimen was ready made. It had three distinct

Many palaeontologists wait a lifetime for 'the golden hammer' when 'you split a piece of rock and, sitting like a kipper laid out on a plate, is the perfect specimen'

lobes, convincing Richard that it was indeed a tri-lobite, and it fitted into the palm of his hand.

'This was my first discovery of the animals that would change my life,' Richard later wrote. 'The long, thin eyes of the trilobite regarded me, and I returned the gaze ... There was a shiver of recognition across five hundred million years.' From then on, he was hooked.

Richard had brought a specimen, about 15cm long, into the studio to be admired. 'This is quite a decent-sized trilobite from Morocco,' he said. 'It's about 400 million years old.' The thorax, being composed of multiple individual segments, was flexible. 'So, this animal, if it was threatened, would have been able to roll up in a ball, rather like a hedgehog or some kind of woodlouse. It has quite a thick exoskeleton made of calcium carbonate, the same as lobsters or many kinds of molluscs, and underneath it has legs, which are very rarely preserved. And antennae, like any other self-respecting arthropod.'

Trilobites inhabited our planet before the super-continent Pangaea, living in oceans that have long since vanished due to plate-tectonic movements and shifting continents. Emerging more than 500 million years ago, as part of the Cambrian evolutionary explosion, they swam around or scuttled on the sea floor, along with jellyfish, clams and snails, when nothing lived on the land, not even plants. They occupied a whole range of ecological niches in the sea, 'which probably accounts for why there are so many different species.' Furthermore, they were adapted to all sorts of different environments, 'living in every water depth from quite shallow to extremely deep' and thriving in temperatures that ranged from tropical to sub-polar. Many were tiny mud-grubbers, but some were trilobite hunters the size of a casserole dish. Some scavenged. Others were filter feeders. Some 'probably ate sediment to extract edible particles'.

'They were extremely successful and diverse for a very long time,' Richard said. 'So when the people I used to meet on my commuter train from Henley on Thames would look slightly surprised that I could spend all day apparently studying one trilobite, I used to have to explain that "Actually, I'm a historian of several hundred millions of years, so there's quite a lot to do!"'

The idea of being highly knowledgeable always appealed to Richard. As a child, he took one of his fossil finds (an ammonite) to the Natural History Museum in London to get it identified and was impressed by the 'rather learned gentleman' who had emerged eventually from behind the scenes to tell him exactly what it was. There and then he thought, 'there was something really rather wonderful about being "an expert".'

He studied geology A Level at school, thanks to Mr Williams, 'a marvellous man'. Williams, a geologist before he became a geography teacher, was delighted to be able to teach a few people like Richard his favourite subject and persuaded the school to embrace this uncommon course.

He went to Cambridge University to read natural sciences, specialising in geology, and during one holiday he worked as a volunteer field assistant on a university expedition to the northernmost part of Norway.

'Spitsbergen didn't enjoy a balmy climate,' he said with a smile. Many days were spent sheltering from blizzards in a small tent, reading Tolstoy and eating porridge. 'That's when I read *War and Peace* and the heavy Russian novels. But I also made wonderful discoveries, including trilobites. By Jove, it was an adventure,' he said, clearly still thrilled.

Weather permitting, Richard and the expedition leader would venture out in a small boat, travelling through iceberg-clogged seas to hunt for fossils. Often, they ended up huddling behind whatever shelter they could find to bash up rocks. Several tonnes of rock got smashed as they searched for evidence of ancient life. 'Only a foolish person tries to dig out the perfect specimen while still in the field,' Richard said. Instead they wrapped up promising finds and sent them home by ship. 'Trilobites are quite robust. Newspaper usually does quite nicely.'

Unwrapping the newspaper parcels back in Cambridge was a second thrill. 'I gradually realised, with mounting excitement, that most of the animals we were looking at had never been seen by the eye of a man before.' They dated back to the Ordovician period (433 to 491 million years ago) and were much older than most of the trilobites that were known about at the time. They were an exciting find.

'I gradually realised, with mounting excitement, that most of the animals we were looking at had never been seen by the eye of a man before'

Richard wanted to do a PhD and (unlike some students, who spend years deciding what to focus their thesis on) for him, it was obvious what it should be about. These Ordovician trilobite finds were hot palaeontological property. 'As luck would have it,' Henry B. Whittington (a world authority on trilobites) was based in Cambridge and he was as excited as Richard by these extraordinary finds. Happily, he agreed to supervise Richard's PhD. 'So, you could say my path was somewhat eased,' Richard said, acknowledging his good fortune.

He went back to Spitsbergen on a second expedition in 1972 with fellow geologist David Bruton, among others. Apparently, Richard turned up 'rather shabbily dressed in an old anorak, a balaclava, some mittens and rubber boots'.

'The Norwegians were well equipped, at least,' David said. They had leather boots, not wellingtons. 'But even this didn't keep the cold out.'

'Well, yes,' Richard admitted, unapologetic. 'My wife tells me that I still turn up slightly inadequately clad, whatever I do.'

Richard and David shared a tent, sleeping in adjacent sleeping bags for 40 nights. David said it was 'a grubby tent' but Richard found Arctic camping, Norwegian style, 'quite luxurious'. There was a large communal tent with salamis and other delicacies hanging from the tent poles, and they would sit up at night and talk around a hurricane lamp. It was all 'delightfully cosy' at 79 degrees North.

Bar one particularly chilly excursion when, David said, 'Richard, with a rucksack full of rocks, slipped quickly but suddenly into the sea.' There were ice floes offshore.

'Oh yes!' said Richard. 'Yes. I don't quite know how I managed that but, erm, I slid down a smooth, polished rocky surface rather elegantly into the Arctic Ocean, where legend has it you have four minutes to get out.'

Fortunately, David noticed and grabbed him with an outstretched arm before his head went under. 'He was freezing and hardly able to walk,' David said. 'Mercifully, he made it back to camp where he dried out, and that night we drank to life.'

On his return from Norway, Richard worked in the Sedgwick Museum in Cambridge and spent most of the next five years chiselling out trilobites from the Spitsbergen rocks and extracting specimens, relentlessly mixing and matching different trilobite bones, assembling head shields and cheekbones, segments of spines and broken legs (on the rare occasions that these delicate limbs had been preserved), trying to re-create the skeletons of these ancient creatures and identify different species.

Working in this way, taking care to avoid any inaccurate misfits, he discovered over a hundred species buried in the Spitsbergen rocks, including several that were new to science. 'This particular place in Spitsbergen' contained a proliferation of different species. 'It was very fortunate,' he said. He described as many of them as he could so that future generations of trilobite hunters who might unearth similar remains would know what they had found, and he named several new species, including *Opipeuter inconnivus* 'the one who gazes without sleeping' (a name that perhaps reveals more about his state of mind at the time, than it did about the trilobite in question).

Then he started to try 'to deduce how these animals actually lived', paying particular attention to their eye sockets. Trilobites have the first well-preserved visual system in the fossil record. 'It's quite amazing how much can be discovered about the way they saw the world through those eyes.'

It had been assumed that all trilobites lived on the sea floor, partly because the specimens that had been found pre-Spitsbergen tended to be flat, and partly because their eyes only seemed capable of looking up. But the specimens that Richard found in Spitsbergen led him to question this assumption. He came across trilobites of all shapes and sizes. Some were quite rounded, and several had huge globular eyes. Eyes that would be capable of looking in every direction, he imagined. Sea-floor dwellers don't need to look down, but for creatures that are surrounded by water, 360-degree vision can be useful. Mainly for this reason, Richard was convinced that these big-eyed trilobites were swimmers. 'And there were thousands of them,' Richard said, excited. 'Like krill.'

The Ordovician ocean, he realised, must have been teeming with swimming trilobites as well as the species he had known about before, which were grubbing around on the bottom. Fortunately, the sedimentary layers in the area of Spitsbergen from which the fossils were extracted clearly alternated between deep and shallow seas, so Richard could cross-reference his anatomical evidence with the environmental clues embedded in each stratum. Were these trilobites found in the kind of rocks that would result from an ancient lagoon or from the bottom of a deep ocean? In this way he was able 'to reconstruct ancient habitats, as well as just naming trilobites'. 'Imagine a sloping shelf, just as we have today at the edge of a continent, perhaps with different communities of trilobites each living at different depths off the coast.'

Meanwhile, in the office next door to his, the eminent structural geologist John Dewey was developing the newish theory of plate tectonics. John was a strong advocate of the idea (developed in the 1950s and 60s) that the outermost layer of the Earth's crust is broken into tectonic plates. These plates are constantly on the move, floating on the treacly liquid beneath, causing continents to drift apart (creating oceans) or to collide (forcing landmasses together and making mountains).

Shifting slowly over deep geological time, these plates change the geography of the world. About 270 million years ago, an assortment of ancient continents (that look nothing like their modern counterparts) came together to form the most recent supercontinent, Pangaea. By studying the structure of the ancient Appalachian Mountains and looking at how the rocks had been deformed, John hoped to gain an insight into the geophysical forces at work in the formation of Pangaea and to determine the shape of the ancient continents that existed before it.

Richard had a different idea. Perhaps trilobites could help to describe these ancient continents? It was 'natural', he said, for him 'to slot his new discoveries from Spitsbergen into a plate-tectonic context'. But perhaps not everyone would have made this leap. By looking at the global distribution of the many and varied species of Ordovician trilobite, Richard found 'a different way of getting at the ancient geography'. All of a sudden, all the careful work he'd done, determining how and where different species of trilobite must have lived, came into its own. His ability to differentiate between deep-sea divers and shallow-sea swimmers (which had been interesting and of value in itself) now provided him with vital clues about the whereabouts of these ancient continents.

He created maps showing the relative abundance of different trilobite species. Wherever there was an abundance of shallow-water species, there must have been shallow seas, and shallow seas, Richard reasoned, are found on the edges of continents (not in the deep oceanic trenches between them). By this straightforward but inspired logic, he was able to map out the boundaries of these ancient continents, using the shallow-water trilobites as his guide. And as everyone who has ever been confronted by a difficult jigsaw knows, the edge pieces can be a good place to start. The existence of deep-water trilobites in the sediment indicated the main body of an ocean; and an absence of trilobites meant that the rocks must have been on land, not under the sea.

Before long, he was presenting papers at international conferences about plate tectonics (a hot topic in the Earth sciences in the 1970s). He escaped the trilobite ghetto and found a platform from which to share his trilobite expertise. Convinced by the power of his trilobite evidence to describe the landmasses that existed before Pangaea, he was determined to show the world just how useful his specialist knowledge could be. And he argued robustly with anyone who had a different picture of the ancient *Mappa Mundi* from the one that the trilobite evidence suggested.

'New results and controversies were happening all the time and getting resolved,' he said. 'We did have several quite big battles between different fields of science, and generally speaking, the palaeontological evidence held up pretty well.'

'We did have several quite big battles between different fields of science, and generally speaking, the palaeontological evidence held up pretty well'

On more than one occasion, and thrillingly for Richard (who had always found physicists a little too sure of themselves and their theories), trilobite evidence trumped measurements of seismic forces. Palaeontology eclipsed physics. His favourite organism was capable of rearranging continents. All power to the trilobites! What more could Richard ask for from the fossil love of his life?

Royal Society of Edinburgh

JOCELYN BELL BURNELL

'I do doggedness not inspiration'

Grew up in: Lurgan, Northern Ireland
Home life: one son, Gavin
Occupation: astronomer
Job title: Visiting Professor, Oxford University
Inspiration: books on astronomy in her father's library
Passion: listening to the universe
Mission: to explain a tiny anomaly in the data from a radio telescope
Best moment: finding a second stellar object that emitted a rapid, regular pulse
Worst moment: 'There were many when trying to combine a career with being a wife and mother of a small child'
Advice to young scientists: 'Hang in there'
Date of broadcast: 25 October 2011

Alerted by a faint, intermittent radio signal coming from space, Jocelyn Bell Burnell contributed to the discovery of a new kind of star: pulsating radio stars. Pulsars are the dense cores of collapsed stars that remain after a supernova explodes. Being dark and extremely dense, they made the idea that there might be black holes in the universe seem more plausible. As a PhD student, she helped to build a new radio telescope designed by her supervisor Anthony Hewish, the Interplanetary Scintillation Array, forging a path through the male-dominated world of physics in the 1960s. When analysing the data, she noticed some tiny 'bits of scruff' and decided to investigate further. Her determination to find an explanation for a 'smudge' in the data that others might have ignored led to a discovery that was worthy of a Nobel Prize.

'Some people call [the 1974 Nobel prize for Physics] the No-Bell Nobel Prize because they feel so strongly, that you, Jocelyn Bell Burnell, should have shared in the award,' Jim said. 'Should you?'

'I think at that time science was perceived as being done by men, senior men, maybe with a whole fleet of minions under them who did their bidding and who weren't expected to think,' Jocelyn said calmly. 'I believe the Nobel Prize committee didn't even know I existed.'

'You don't sound bitter about this,' Jim said. Then he played Jocelyn a clip of her supervisor Antony Hewish saying, 'My analogy really is a little bit like when you plan a voyage of discovery, and somebody up the mast says, "Land ho." That's great, but who actually inspired it and conceived it and decided what to do? I mean, there is a difference between skipper and crew.'

'That's Tony and the way he feels. That's OK,' Jocelyn replied, untroubled.

'So a shrug of the shoulders is still the appropriate—'

'Yes. It's not worth getting het up about.'

'Many people were and still are outraged on your behalf,' Jim said, not quite ready to let it go. 'Including a former secretary of the Nobel Prize committee, Anders Barany.' Cambridge Professor of Astronomy Craig MacKay said, 'I do feel that Jocelyn did make an extremely important discovery. She's had many prizes and awards over the years, but an awful lot of people really do feel that this wasn't particularly fair.'

'The world's not fair and it's how you cope with the world's unfairnesses that counts'

'The world's not fair,' Jocelyn said, 'and it's how you cope with the world's unfairnesses that counts.'

Jocelyn encountered discrimination early in life, but her parents were outspoken about it. She grew up in Lurgan, Northern Ireland. In her first week at Lurgan College Preparatory Department in September 1948, a message went around the class: girls go to one room, boys to another. 'The girls were sent to the domestic science room and the boys to the science lab. My parents read the riot act when they heard that,' Jocelyn said. 'So did the parents of one or two other girls. And the next time the science class met there were three girls and all the boys.'

From the age of 11 she attended an all girls' Quaker boarding school, where there was no such prejudice. The physics teacher at the Mount School in York insisted that 'physics was easy' and encouraged all the girls. Arriving at Glasgow University in 1961, however, she was greeted by wolf-whistles and the thunderous noise of her fellow students stamping their feet when she entered the lecture hall. She was the only woman in her year doing an honours degree in physics, and this pattern of behaviour continued for four years.

'It was the tradition in Glasgow at the time that whenever a woman entered a lecture theatre all the blokes did that,' she said.

'Goodness me,' Jim muttered under his breath.

'Mmm, building site,' Jocelyn replied. 'I was a bit

annoyed with the staff, the faculty, who must have known what was going on . . .'

'It's not been an easy ride for you, has it?' Jim said. 'As a woman pursuing a scientific career that began in the 1960s?'

'No, it's not been easy . . . Women of my generation who've stayed in science have done it by playing the men at their own game. And I do wonder a bit what it's done to me as a woman. Am I now a she-male as a result of forty years of that?'

'Are things easier for women in science now?' Jim asked

'I think it is easier. It could perhaps be easier still, but it's much, much improved. That's undoubtedly the case . . . Curiously, there is a higher proportion of women in astronomy than there is in physics. And given that most of us have come through physics to get there, I find that particularly interesting . . . Perhaps it's slightly more cooperative, less competitive. Perhaps it's more open to different ways of thinking, different approaches, than the rest of physics. I'm not sure.'

Undeterred by the sexism she had encountered, and enthused by her physics lessons at secondary school, Jocelyn developed an interest in astronomy, reading her father's library books. 'But the snag about conventional astronomy is you do it at night, and I wasn't really very keen on that,' she said.

'For a while I thought, "I'm not an astronomer." Then I discovered all these new astronomies which had grown up post-Second World War.' Both optical and radio astronomers detect electro-magnetic radiation. Optical telescopes look for visible light, itself comprised of a rainbow of colours from violet light (which has a wavelength of about 400 nanometres) to red light (wavelength about 700 nanometres). Radio telescopes search for much longer wavelengths. And just as a radio tuned in Britain to 92–95 FM will pick up BBC Radio 4 and not some other radio station, so different telescopes only detect the particular wavelengths they are designed to find within the spectrum of electro-magnetic radiation.

'To look up into the sky and look only at the visible parts of the spectrum is a bit like listening to a piano piece by Beethoven and only being able to hear the middle octave,' said Jim. 'Radio astronomy opens up the rest of the piano.'

'And it's a huge scale indeed,' said Jocelyn.

When Jocelyn was a student, 'radio astronomy was very new. It was really going places.' And it provided an added bonus: there was no need to lose any sleep. To detect light from elsewhere in the universe, it needs to be dark here on earth. Radio waves from space are detectable at any time of the day (or night). They are, however, very faint. Imagine adding up all the energy from all the extraterrestrial radio signals ever received, excluding the radio waves generated by our sun. The total energy from all those radio waves would not melt a snowflake.

Karl Janksy, a physicist working at Bell Laboratories in the US in the early 1930s, discovered cosmic radio waves when he was trying to locate the source of some unwanted interference. Many astronomers were underwhelmed by his discovery, but he was enthused. He built a radio telescope, which came to be known as 'Jansky's Merry Go Round' on account of its rotating antenna, and published the first radio map of the universe in 1933.

Radio astronomy was then transformed by the radar

technology that had been developed during the Second World War. Armed with two trolleys of ex-Army radar equipment, the British physicist Bernard Lovell built a dish, 76m wide, to catch as many radio waves as possible and moved his Mark I telescope to the Cheshire countryside when the signals from electric trams in Manchester started causing unwanted interference.

It took six years to build, and the budget increased tenfold as the work progressed. Critics accused him of misusing public funds, and the Public Accounts Committee investigated. But when Lovell's telescope was able to track Sputnik 1, the satellite launched by the Soviet Union in October 1957, they were reassured that Lovell's 'Big Ear' in the middle of a field was not the expensive white elephant they feared.[1]

Lovell's primary purpose, however, was radio astronomy, not winning the Cold War, and Jocelyn learnt about the exciting discoveries that were being made. Using his telescope, she worked as a summer student at Jodrell Bank, 'which was huge fun'. The results from the telescope were meant to be analysed by a computer in Manchester, but the landline which connected the telescope to the computer was down, so Jocelyn and the other summer students were employed as human calculators, 'basically to do long division. It fairly improved my arithmetic that summer,' she said, smiling. 'The lecturers back in Glasgow commented on it afterwards.'

Inspired by the experience, she applied to do a PhD at Jodrell Bank, but the graduate students told her they

---

1 Later it tracked the Eagle lander as it descended to the surface of the Moon in 1969, carrying Buzz Aldrin and Neil Armstrong, firmly embedding itself in the national and international psyche as an excellent scientific resource.

didn't rate her chances. '[Lovell] won't take a woman,' they warned. 'There was a woman student once,' Jocelyn was told. Apparently, 'she and a male colleague had put the dormitory to a use for which it wasn't intended.' The boy had boasted about his conquest. 'Sir Bernard got to hear about and said, "No women."'

'It's a very fair world this, isn't it?' Jocelyn said. 'No women.'

'And that was it. Your chance of working at this brand-new wonderful facility was gone,' Jim said, horrified.

'I always suspected that I wouldn't get to work at Jodrell Bank,' Jocelyn said. 'There was, of course, Cambridge, but the standard there was phenomenally high.' Instead, her plan was to go to Australia, 'where they did a lot of radio astronomy and did it very well'. However, she had a few months in hand before the Australian academic year began. 'So I applied to Cambridge just in case, and to my very great surprise got in. I think partly because I had my own money,' Jocelyn said. 'But it was a huge surprise, and I did find the place quite daunting. Have you heard of imposter syndrome?'

'No,' Jim replied.

'It's something that afflicts people who lack confidence, women often and some men. They find themselves in a place they didn't expect to be, somewhere prestigious, and they say to themselves, "Ooh, they've made a mistake admitting me here. They're going to throw me out. It's all their fault. It's a mistake. I shouldn't be here. I should leave before they find me out."'

Jocelyn joined the new Radio Astronomy Department in Cambridge led by Martin Ryle in 1965. Her supervisor

Antony Hewish had designed a new telescope, the Interplanetary Scintillation Array, hoping to detect quasars. Twenty of these extremely luminous celestial objects had been discovered by radio telescopes, and astronomers were keen to find more.

'It was a huge telescope' made up of poles and wires spread over an area 'the size of 57 tennis courts'. 'I spent the first two years building the equipment, along with about half a dozen other people. It was pretty heavy physical work,' Jocelyn said. Her responsibility was to connect cables in various ways and plugging things in. 'One hundred and twenty miles of wire and cable were kept up out of the wet grass by being strung on posts and the posts had to be sledgehammered into the ground . . . By the end of my PhD I could swing a sledge. It's not something I expected to do, but I could do it.'

'Most unusually, when we switched that radio telescope on, it worked first time,' Jocelyn said. 'They don't normally do that, but ours did. We had a very careful technician working with us, Don Rolf, and it's actually thanks to him . . . that phase [of the project] was particularly successful.'

Once the telescope was up and running, the more sedentary activity of data analysis could begin. 'As it scanned the cosmos for radio waves, the data generated was automatically recorded on a paper chart, which exited the machine at a rate of 100ft a day.'

'I ran it for six months, which gave me about three miles of paper, which I had to analyse by eye.' Jocelyn would go through the charts logging anything that she thought might be a quasar. She also identified chunks that looked as if they were generated by man-made devices. The telescope was located in the countryside, so as to be as far away as possible from electrical devices, but there was still some radio interference from non-cosmic sources.

Occasionally, however, there was a strange additional signal that didn't look like either a quasar or low-level interference from man-made machines. It appeared maybe 'One time out of five, or one time out of 10 when we looked at a particular bit of sky. And it occupied about a quarter-inch of the chart', in a weekly chart that was 700ft long.

'That's a tiny detail that most mere mortals presumably would have overlooked,' Jim said. 'You found it interesting enough to follow up. Why?'

'Because I was being extremely thorough, conscientious, pedantic and wanting to make sure I understood how the equipment was behaving,' Jocelyn said. 'And this was a signal I couldn't explain, so it troubled me.'

'Were you also maybe a little bit worried that you'd made a mistake or that something was wrong with the telescope or?'

'Yes. One of the reasons for checking out all the signals you get is to make sure the telescope is functioning as you expect it to, so that's important. I'd been responsible for the wiring, so I was anxious about crossed wires and things like that. The first time I saw the signal I just logged it with a question mark. But after I realised it was reappearing from the same bit of sky, I showed it to Tony and said, "What do you think we do about this?"'

'What was his response to seeing it?' Jim asked.

'The signal was all crammed into this quarter-inch stretch of chart. It really was jammed together, and you couldn't see what was happening. So we decided that really what we had to do was to spread it out a bit, which means running the chart paper faster underneath the pen. And I did that for a month. And absolutely nothing showed.

Tony got very cross: "Oh, it was a flare star, and it's been and gone and you've missed it," he said, that kind of thing.'

Jocelyn persisted, unwilling to dismiss what she had seen quite so readily. She knew the strange 'bits of scruff' from previous observations were evidence of something and they remained unexplained. What was going on to create these tiny, blurry smudges in the data that popped up every so often? She continued to run the paper faster under the pen, hoping the additional signals would reappear and that she would be able to analyse them in more detail.

'And then finally it came back, and I got it,' she said, reliving the excitement. 'And it came in as this string of pulses, equally spaced pulses.'

'I have a photocopy of your original data here,' Jim said. 'It's amazing, even for the untrained eye. It's clearly something that's oscillating at regular intervals . . .'

'And that's what disturbed Tony,' Jocelyn said. The rapidity of the oscillations. They had a period of one and a third seconds, which suggested the signal was coming from an object that was very small. 'Tony knew this,' Jocelyn said. 'I didn't.'

Jocelyn phoned him up that afternoon and said, 'This scruff is a string of pulses one and a third seconds apart.'

Such a regular and deliberate signal was most likely to have been generated here on Earth. Jocelyn suspected otherwise. Her calculations suggested that the signal was generated 200 million light years away, 'which is well out in the galaxy, beyond the solar system'. Together they joked about it being produced by little green men and called it LGM-1 for fun.

'What did you do to try and confirm that this was really coming from space?' Jim asked

'It was a very anxious time for me,' Jocelyn said. 'We spent about a month doing test after test after test and finally established it wasn't a fault with the equipment.' Confident that it was 'stellar not terrestrial', she could describe the different characteristics of this unusual source of radio waves but still had no idea what it was, 'because nobody had seen anything like this before'.

'It was a very rapid pulse. And it was a very accurate pulse. Because it was rapid, we could put a size limit on the thing – probably about 20,000 miles across.'

'Which is much smaller than a star,' Jim said.

'Yes. Much, much smaller than a star. A big planet maybe, but much, much smaller than a star. It was also surprisingly accurate. It kept its pulse period. It didn't get tired. It didn't flag. And if something's going to keep pulsing very, very accurately, it's got to have great reserves of energy, so it has to be big.' As Einstein's famous equation $E= mc^2$ (where E = energy, m = mass and c = the speed of light, which does not change) tells us, the more massive an object is, the more energy it contains.

'At the time we couldn't get our heads round it,' Jocelyn said. 'The regular unflagging pulse suggested a massive object. The speed of the oscillations suggested something very small.'

'It was small in size and very heavy, so you're saying therefore that it was a very dense object?' Jim said

'That's what emerged ultimately,' Jocelyn said. This conclusion was far from obvious at the time. It had been suggested that there must be such a thing as black holes in the universe in 1930s, but very few astronomers took the idea seriously. Pulsars are 'preposterously dense'. 'There was nothing like them,' Jocelyn said.

'So dense that it's like getting all the people on Earth squeezed into a thimble,' Jim said.

'And there's some very interesting physics, because it's closely packed.' But at the time, Jocelyn and her colleagues 'just couldn't envisage what was going on.'

One swallow does not a summer make, and so it is with strange astronomical objects.

'Finding the second pulsing object in the sky, was that the moment more than any other where you felt this is it, I've found something really new?'

'That was a really sweet moment, because that's when you become pretty convinced you've stumbled over a new kind of distribution, a new population, whatever.'

If there were two of these strange pulsing objects, then it seemed highly possible that there might be more. 'And it was a particularly difficult observation to do, getting that second one, so it was quite a technical achievement as well.' This second observation ruled out the possibility, however remote, that the signal was caused by little green men or another rogue source, and pointed instead to the discovery of a new type of astronomical object. Jocelyn and Anthony wrote a paper describing the discovery of these pulsating radio stars, or pulsars.

'And on 24 February 1968 you were the second author on this key paper in a top journal, in *Nature*,' Jim said.

'We should pause just for the benefit of listeners,' Jocelyn said. 'One of my fellow students at Cambridge published a paper in *Nature*, and his mother was terribly upset. She thought it was the magazine of a nudist colony! It's a prestigious scientific journal – the top one, folks.'

'And the *Daily Mail* reported "Girl Discovers Little Green Men?"'

'Oh, yes, and worse than that, what were my vital statistics and how tall was I and, you know, chest—'

'Good grief,' Jim interjected.

'Waist and hip measurements, please, and all that kind of thing. The press didn't know what to do with a young female scientist. You were a young female. You were page three. You weren't a scientist. What sticks in my mind is the difficulty I had in dealing with those reporters. I would just love to have been really rude to them, but I was a PhD student. I was dependent on senior colleagues for references. I wasn't in a position to offend the press, because the lab needed the publicity.'

'The press didn't know what to do with a young female scientist. You were a young female. You were page three. You weren't a scientist'

'Is there anything that you think you could have done or said at the time that would have made things easier for you?' Jim asked

'I don't think so. If there had been a senior woman around, she might have been in a strong enough position to weigh in and say to the press, "Look, these are irrelevant questions," or something like that, but there wasn't.'

Six years after the *Nature* paper was published, Anthony Hewish and Martin Ryle, the head of department, were jointly awarded the 1974 Nobel Prize for Physics for the discovery of pulsars. The Nobel Committee allows up to three people to share an award, but Jocelyn was not included.

'Was there a point when you started to think that you deserved some of the credit?' Jim asked.

'I think there were probably a lot of moments like that during the phase where I was moving around after my husband and working part-time trying to keep in the field. It was in general a very tough time.'

Jim wondered whether Jocelyn's humility had held her back: 'Your colleague Craig MacKay says, "Jocelyn does meek."'

'I think that's a misjudgement,' Jocelyn said, firmly.

'I think people fail to recognise the grit that was needed just to be there. To have gone through a degree in Glasgow as the only female with the treatment I had, to get myself to Cambridge, to hold my own . . . I had similar problems when I was trying to combine a family, career, and my colleagues thought I was not serious because I was working part-time, totally failing to recognise the doggedness needed to get working at all in Britain at that time when you were a mother.'

The discovery of pulsars changed the way we see the universe. Previously, it was widely assumed that stars exploded catastrophically at the end of their lives, leaving nothing behind. Those spectacular explosions, or supernovas, were thought to result in the complete annihilation of the stars. 'There were one or two mad astronomers around who suggested otherwise, but they weren't part of the mainstream.'

Pulsars, we now know, are 'the end points of the life of massive stars. The core of the original star gets very compressed, very shrunk, during the explosion, and that's what makes these very dense, small stars . . . Things that people had dreamed about, and we thought were really not terribly sensible, turned out to be real.' And when

'The core of the original star gets very compressed, very shrunk, during the explosion, and that's what makes these very dense, small stars'

the existence of objects that were so unimaginably dense became clear, the idea of there being black holes in the universe seemed a little less preposterous. 'They certainly made black holes much more credible,' Jocelyn said.

'Did you realise at the time, back then in the late 1960s, the significance of the discovery?' Jim asked.

'I don't think any of us did, to be honest,' Jocelyn said. 'It was only over the next few years when these neutron stars, which are what pulsars actually are, were found by other astronomers, in X-ray astronomy for instance, that we began to realise the magnitude of the discovery we'd made.'

Nearly 2,000 pulsars have been discovered since Jocelyn first noticed those 'bits of scruff'. 'We're a bit limited by the present equipment, because pulsars are pretty weak,' she said. 'We can see the ones in the near half of the galaxy. The far half of the galaxy is pushing it, for example. So there's a lot more to be found.'

It remains a hugely dynamic field of research 'with staggering results rolling in each year'. The American scientists Russell Hulse and Joseph Taylor were awarded the1993 Nobel Prize for Physics for discovering the first binary pulsar (PSR 1913 +16), and they invited Jocelyn to Stockholm to be their guest of honour. Jocelyn missed out on a Nobel Prize but has been showered with other prizes ever since. 'I've had so many other awards, it's been amazing' she said. 'And it still goes on and on and on.'

Oxford University Images/Joby Sessions

MARCUS DU SAUTOY

'Doing mathematics is a bit like being a Buddhist monk'

Grew up in: Henley-on-Thames
Home life: married to Shani, a psychologist. They have three children: Tomer, Magaly and Ina
Occupation: mathematician
Job Title: Professor of Mathematics, New College, Oxford and Simonyi Professor for the Public Understanding of Science
Inspiration: his teacher Mr Bailson, who encouraged him to find out what mathematics was really about
Passion: the mathematics of symmetry (Group Theory)
Mission: to predict how many symmetrical objects could possibly exist
Best moment: discovering a new symmetrical object
Worst moment: a graduate student disproving a conjecture that he had been working on for ten years
Advice to young mathematicians: fail, fail again, fail better
Date of broadcast: 7 June 2016

Marcus du Sautoy is master of an abstract algebra known as Group Theory, the mathematical language that describes symmetry. He shot to fame, of the mathematical kind, aged 24, when he found a pattern that helps to predict how may symmetrical objects are logically possible for a given number of symmetries. In 2001 he won the prestigious Berwick Prize for the best mathematical research by a mathematician under 40.

An invitation from an editor at *The Times* launched his career as a communicator of science and mathematics that now runs in parallel with his mathematical research. He is the Simonyi Professor for the Public Understanding of Science, having taken over from Richard Dawkins in 2008. In his own play *X&Y*, he played the character X. And he has given talks at the Royal Opera House about the mathematics in Mozart's *Magic Flute*. 'It's easy to preach to the converted,' Marcus says. He'd rather tackle the opposition.

'I started learning the trumpet at exactly the same time as I fell in love with mathematics,' Marcus said. 'They are both very abstract worlds where you can create structures.'

'And find patterns?' Jim suggested.

'Patterns. Exactly! Patterns,' Marcus said, highly animated. He gloriously defies crass stereotypes of socially awkward mathematicians and is fantastically engaging and energetic about his subject. 'A mathematician is a pattern searcher. That's what we do all day. We try to find patterns in the chaotic world around us.'

'A mathematician is a pattern searcher. That's what we do all day. We try to find patterns in the chaotic world around us'

Sitting in his study in Stoke Newington, north London, often with three children playing not far away, he lets the world around him disappear, allowing himself 'to get sucked into a very abstract world'. It's a form of meditation, 'a bit like being a Buddhist monk'. He imagines mathematical worlds that 'somehow can't exist in the real world'. 'I am interested in all kinds of different possibilities. Not only what is real.'

Then the challenge is to maintain a meditative focus so as to explore these new domains and discover new mathematical terrain. It involves a lot of playing, a lot of trying, and a lot of failing. It is as Samuel Beckett said: 'Ever tried. Ever failed. No matter. Try again. Fail again. Fail better.' He has a study in the mathematics department in Oxford but finds working at home takes the pressure off. He feels less obliged to deliver results. Listening to music also helps to settle the nerves, and when he's not meditating on mathematics, he scribbles on yellow legal pads. 'Somehow the colour yellow works for me,' he said.[1]

'Sometimes, something works.' Occasionally he enjoys a moment of revelation, a moment of 'Oh, now I see that.' And it's the memory of those moments that sustains him. 'That's what I kind of live off,' he said, 'that drug.'

'There's a terrible tendency, I think, to imagine that mathematicians are rather unemotional creatures . . .' Jim said.

Marcus agreed. 'When people think, "What is a mathematician actually doing all day?" they tend to think I might just be proving all the true statements about mathematics. That's not true. I'm making a lot of choices and those choices about what becomes part of mathematics are actually steered by emotional reactions.'

'Logic and emotion aren't necessarily mutually exclusive?' Jim asked.

'No. That's right. And I think the idea of proof actually taps into those two ideas, *logos* and *pathos*. A mathematical proof has got to be logical. But it also has to have pathos. It's got to move you emotionally.'

'It does seem to be this strange mix of being absolutely logical and rigorous and at the same time, highly intuitive and creative,' Jim said.

'Absolutely!' Marcus agreed. 'You have to have intuition about where you are going. Often, you'll see the destination way before you know how you're going to get there. That's what a conjecture is. You're saying, I think

1 Perhaps because the mathematical journals that introduced him to the subject as a teenager all had yellow covers.

this is going to be true, and then you apply your logical mind to try and find a path to that distant mountain. It's a very creative subject.' It's also about proving that what you have imagined represents a mathematical truth, not a flight of fancy, by writing rigorous proofs based on a series of indisputably logical steps.

'So is mathematics, for you, more of an art than a science?' Jim asked.

'I think I chose mathematics because it bridges both. It's certainly the language of science and you really can only do science if you turn it into a mathematical language, I think. But on the other hand, there is this creative side where you use your imagination to create worlds that couldn't exist in the physical universe.'

'You use your imagination to create worlds that couldn't exist in the physical universe'

As a child, Marcus was good at mathematics but not exceptional. He enjoyed playing the trumpet, was very keen on languages and wanted to be a spy. He didn't enjoy learning his times tables and there were several others in his class who seemed to him to be more able mathematically than he was.

'It was really my mathematics teacher that helped me to discover this world,' Marcus said. 'I was at my local comprehensive school and in the middle of the class my mathematics teacher said, "Du Sautoy! What are you doing after class?"' Marcus assumed he was in trouble for talking or some other misdemeanour. But rather than issue a detention, Mr Bailson suggested that Marcus should find out more about the big ideas in mathematics. 'It's not what we're doing in the classroom,' he said. 'It's not the technical side. It's not long division, percentages, sines and cosines.'

Many years later, he asked Mr Bailson why he had singled him out. 'Was it a random experiment?' he asked, still not convinced that his abilities at school were worthy of any special attention. No, it was not. Apparently Mr Bailson had not done such a thing before or since. He said he could see Marcus responding to and enjoying abstract thinking and thought he might find mathematics delightful. He was right. 'He gave me the key to the Secret Garden of mathematics,' Marcus said. 'And that's when I realised, "Hey, this is a wonderful world. This is where I want to spend my time playing."'

His father, normally an unemotional man, cried when Marcus told him that he'd been awarded a scholarship to read mathematics at Wadham College, Oxford. Reading the mathematical journals that Mr Bailson had recommended, he had learnt about all the exciting developments in the study of symmetry in the early 1980s. 'The language I fell in love with, actually right from school age, was something called Group Theory,' he said. 'It's the language to understand the world of symmetry and it's a lovely language because it turns geometry into algebra.'

'And just to be clear,' Jim chipped in, 'when you talk about symmetrical objects, you're not talking about solid things in real space, you mean hypothetical, more abstract structures that could exist in three or four or five or a million dimensions?'

'I think that's one of the exciting things for me,

mathematically,' Marcus said as he spun off into an enthusiastic explanation of symmetry. 'The discovery of symmetry started with the dice that was used by the ancient Egyptians.' For a dice to be fair there needs to be an equal chance of it landing on any one of the faces, and so it needs to be perfectly symmetrical. Plato later identified 5 perfectly symmetrical solids: the triangular pyramid or tetrahedra (4 faces), the hexahedra (6 faces like a cube), the octahedra (8 faces), the dodecahedra (12 faces) and the icosahedra (20 faces). All these Platonic solids make good dice.

'Then we realised that symmetry is something much more abstract,' Marcus said. 'We realised that you can create dice in higher dimensions . . . This is the powerful language created by René Descartes . . . Any geometric object can be changed into a set of Cartesian co-ordinates.' Any location on the surface of the Earth can be described by two numbers, longitude and latitude. To define a point in three-dimensional space, three co-ordinates are required. If you are working in four dimensions, four co-ordinates are needed. And so on.

Using numbers to describe shapes enables us to move beyond three dimensions. 'Even though geometry runs out in three dimensions (or maybe four, by adding time or something), the co-ordinates side of the dictionary goes on.' Using the language of co-ordinates, 'I can start creating cubes in four-, five-, six-, infinite-dimensional space,' Marcus said.

Most of us think in three dimensions, or perhaps four if we imagine three-dimensional objects moving through space, over time. But beyond three or four dimensions, our everyday imagination fails. A mathematician, however, just adds another set of co-ordinates. Any point on a two-dimensional piece of paper is defined by two co-ordinates, traditionally described as (x, y). In three-dimensional space, three co-ordinates (x, y, z) are required. In four dimensions, four co-ordinates. And so on.

'I don't have a geometric mind. I don't view mathematics through pictures, as some other people do. I love manipulating language.'

In the early 1980s Group Theory was an exciting field. The study of symmetry was coming together and the Group Theorists were on a roll. A group is an abstract idea. It's a set of elements (numbers, for example), together with a way of combining them (such as addition or multiplication) in which all members satisfy certain conditions. Principally, if you combine two elements in the group, the result is also in the group. One special class of groups, the so-called simple groups, cannot be broken down into smaller pieces and are sometimes called the atoms of symmetry for this reason.

John Conway, a man who Marcus likes to call the Long John Silver of mathematics, had decided to create a reference book describing all the 'atoms of symmetry', just as Dimitri Mendeleev had classified all the elements in a periodic table, leaving gaps for elements that logically ought to exist but had not yet been discovered. And mathematicians were hunting frantically for ever larger simple groups In 1973 'the monster' was discovered, a simple group that exists in 196,883 dimensions. Things had moved on from the Platonic solids. Conway et al.'s *Atlas of Finite Groups* was published in 1985. It was an extraordinary reference book that described and classified all the atoms of symmetry that had been found: 94 pages packed with mathematical charts depicting islands and archipelagos revealing the location of these symmetrical treasures. It was a guide to an abstract world of symmetrical possibilities in hyperspace, the culmination of 150 years of mathematical research.

Later that same year, Marcus went to visit John Conway and his group in Cambridge, thinking it might be a good place to study for a PhD. He remembers John producing the *Atlas of Finite Groups* and slamming it on the table.

'What's your name?' John demanded, having said little else. 'And your initials?'

'M. P. F. du Sautoy,' Marcus replied, a little bewildered.

'The surname is fine but the "du" will have to go, and so will one initial,' Conway said abruptly. The reason soon became clear: all the *Atlas* authors had six-letter surnames and two initials:

J. H. Conway
R. T. Curtis
S. P. Norton
R. A. Parker
R. A. Wilson

If Marcus were to keep his full name, it would break the symmetry. It would also, apparently, have been necessary for him to spell his name Zautoy, not Sautoy, because all the authors to date had joined this group of Group Theorists in alphabetical order.

All things considered, Marcus decided to stay in Oxford for his PhD, and not because he didn't want to change his surname. He was worried that the Cambridge Group Theorists had done what needed to be done. 'Perhaps this area had been finished with this classification and there was nothing new to discover.' Maybe things had come 'to a natural end'. 'So I started a new project,' he said.

If the *Atlas* described the atoms of symmetry, Marcus wondered what 'molecules' could be created when you started putting these atoms together. 'What could you make with these building blocks? Is there a pattern which will help us to predict how many symmetrical objects we'll get as the number of symmetries of that object increases?'

'I have to ask you this,' said Jim, a little sheepish. 'Why do you want to know how to predict the next symmetrical object?'

'Symmetry is one of the fundamental languages of the universe,' Marcus said, welcoming the question. 'You have symmetry bubbling under physics. It helps to explain how the fundamental particles behave. It describes how a virus will work. So, symmetry is extremely important in our universe. If I can understand symmetry, then I might understand new things about the universe.

'But then symmetry starts to take on a magic of its own. And you forget the universe and say, "This is just amazing! What symmetries are possible?" I'm interested in all kinds of different possibilities, not only what is real . . . That's the fun thing, creating worlds that are impossible physically but possible in the mind.

'I think the beauty of mathematics, and very often the motivation for mathematicians, is just the beauty of creating new knowledge and the fact that there's a sort of immortality to it as well. Once you create a piece of mathematics, it doesn't get overturned.'

Unlike scientific knowledge, which is always provisional,

a mathematical truth lasts for all time. 'The power of proof is why I was drawn to mathematics. That kind of certainty. The proof gives me this thing that will remain there for ever. It is very intoxicating.'

'The power of proof is why I was drawn to mathematics. That kind of certainty. The proof gives me this thing that will remain there for ever. It is very intoxicating'

Before Marcus could write a proof, he needed something to prove. What he wanted to be able to predict was this: how many symmetrical objects are there that have a given number of symmetries. And if you change the number of symmetries, from say 27 (3^3) symmetries to 81 (3^4), how does the number of objects in that category change?

For a long time, he searched for a pattern, suspecting a Fibonacci-type rule. (Fibonacci numbers are created by adding the two previous numbers together to get the next number in the series. So the sequence goes: 1, 1, 2, 3, 5, 8, 13 and so on.) But he found no apparent order. No structure. No joy. Several years later, however, several strands of his thinking started to come together and a pattern emerged. And filling multiple yellow legal pads with notes explaining his mathematical meditations, he started to develop a proof.

'I understood that there's a Fibonacci-type rule which helps us to predict how many symmetries any given object will possess. It actually revealed a strong pattern in what, at first sight, looked really wild and random.'

His DPhil thesis (the Oxford name for a PhD), 'Discrete Groups, Analytic Groups and the Poincaré Series', was published in 1989 and reprinted in the *Annals of Mathematics*, a highly exclusive journal that is home to some of the great mathematical works. Marcus was then invited to present a paper at the International Congress of Mathematicians, an illustrious meeting that takes place once every four years. It's where the global mathematical elite gather and the Fields Medal, the mathematical equivalent of a Nobel Prize, is awarded. 'It's what every mathematician aspires to,' Marcus said.

He presented his paper, bursting with pride and joy, only to realise that most of the world-class mathematicians at this highly exclusive meeting could only admire his work from a distance. Many of his more elegant mathematical manoeuvres were missed by mathematicians who were unfamiliar with the mathematical language he was using. Intricate logical constructions were overlooked. 'There were probably only ten people in the world who appreciated the nuances of what I did in my DPhil thesis,' he said.

Doing mathematics is a lonely endeavour. 'All the work I've done since has felt like that,' Marcus said, subdued. 'That's a frustration. You want more people to understand the thrill and excitement of that moment of discovery.'

Then came the terrible challenge, familiar to anyone who has created a masterpiece: what to do next? Jim wondered if Marcus had suffered at all from blank-canvas syndrome, in the wake of his much-celebrated proof: 'Is there pressure? Self-imposed pressure? Did you worry that maybe you wouldn't be able to achieve the same greatness again?'

'That was a very a stressful period,' Marcus said, reflective. None of his new ideas seemed to be working and he 'contemplated running away with a theatre company'. He had done a lot of community theatre work when he was at school and had enjoyed it.

Instead he got a post-doc job at the Hebrew University of Jerusalem and was rescued from the intellectual wilderness by a senior Israeli mathematician, Avinoam Mann. Mathematical insights come in waves, he reminded Marcus: 'There are peaks and troughs.' 'There is a pattern,' he said, appealing to him as only a fellow mathematician could. The mathematical data on mathematical success predicts there will be another peak. You just have to have faith.

He kept going through the fog and, sure enough, there were more mathematical peaks. In 2001 Marcus was awarded the prestigious Berwick Prize for the best research by a mathematician under 40. By this time, he was a Professor of Mathematics at Oxford and later he became a Fellow of All Souls College. At a Fellows' dinner one night, the man sitting next to him asked Marcus what he did. 'Oh, that's so sexy!' the man said. 'Write me an article!' He was an editor at *The Times*. Marcus was not convinced: 'I didn't want to put my head above the parapet.' Presenting proofs to the best mathematicians in the world was one thing: writing a newspaper article, quite another.

Three years later at another Fellows' dinner, some of the Fellows had changed but many of the guests, and the seating plan, remained the same. 'You never wrote me that article,' the editor reminded him. Marcus was amazed and embarrassed that he had remembered. This time he said yes.

It was a steep learning curve. 'That one newspaper article took me a month to write,' he said. It was, nonetheless, the beginning of a marvellous media career, which extends way beyond traditional print media, radio and TV to embrace music, dance and theatre. He has helped to create mathematically enlightened choreography and performs on stage himself, dancing and acting out mathematical ideas. In one of his own plays, *X&Y*, he played the character X.

Midway though his career, he wanted to give something back to the community that had nourished him as a young man. He remembered hearing Christopher Zeeman talk about the Poincaré Conjecture at the Royal Institution Christmas Lectures when he was 13. If one of the greatest mathematicians of the twentieth century had found time to talk to the next generation of mathematically curious young people, surely he could and should do the same. When nearly three decades later Marcus delivered his own Christmas Lectures, 'The Num8er My5teries', he made sure to thank Christopher Zeeman.

Around this time, scientists were being roundly criticised for failing to engage with society. In the Jenkins Report to the House of Lords in the year 2000 they were accused of hiding in their ivory towers and held partly responsible for the breakdown in trust that followed the BSE crisis and the backlash against GM crops. The Royal Society took the criticism seriously and actively encouraged scientists to engage with the public. Marcus, Brian Cox and several others rose to the challenge.

Jim suggested that Marcus's style of science communication was more arty than most. 'You embrace all that luvvie stuff,' he joked.

'Ah yes. I've gone to the dark side,' Marcus replied, with a hearty laugh. 'I think actually it's about finding innovative ways to try and communicate scientific ideas.' Writing and talking about mathematics are just a fraction of the different ways he's tried. He often uses music as a way in. 'Music has always been a powerful way to reveal what mathematics might be about.'

In a series of highly interactive performance lectures on Mozart's *The Magic Flute* at the Royal Opera House, five singers and a pianist searched out patterns, pointing out twists and turns, making connections between musical phrases. And he was delighted when people came out saying they didn't realise there was so much maths in Mozart's music.

Pulled in by a desire to conquer the unknown, to prove what is true and will always be true, Marcus has visited worlds that are beyond the imagination and ingenuity of all but a handful of mathematicians. He has found a way to, and through, highly abstract worlds. He navigates the possible, not the real, and in so doing he has, perhaps, travelled further than most of the explorers in this volume, without leaving his study in Stoke Newington, North London. He remains a master of Group Theory, but the answer to one rather simple-sounding question remains elusive. How many symmetrical objects are there?

'I do not want to die without knowing what the answer to that is,' he said, which felt like either a good place to end, or an exciting place to start.

Courtesy of Chris Lintott

CHRIS LINTOTT

'I take the "Citizen" in Citizen Science very seriously'

Grew up in: Devon
Occupation: astronomer
Job Title: Professor of Astrophysics, University of Oxford
Inspiration: a summer spent camped out at Torquay Boys' Grammar School Observatory
Passion: galaxies
Mission: to recruit citizen scientists
Best moment: being in charge of the James Clerk Maxwell telescope
Worst moment: breaking the James Clerk Maxwell telescope
Advice to young scientists: 'It's ok not to know'
Date of broadcast: 17 January 2014

It was the broadcaster and astronomer Patrick Moore who first opened Chris Lintott's eyes to the wonders of the universe. He made his first appearance on the BBC TV show *The Sky at Night* when he was just 18, and presented the show with Patrick for many years, a job that he continues to enjoy.

His PhD thesis on how stars form used data from the most ambitious astronomical survey to date, the Sloan Digital Skies Survey. But before he could get started, he needed to sort a million galaxies according to their shape – elliptical or spiral – and no amount of digital technology could help. Overwhelmed by the scale of the task, he met up with a fellow PhD student, Kevin Schawinski, and over a pint in the Royal Oak in Oxford, they came up with an ingenious solution. Delegate on an unprecedented scale by inviting everyone to help.

When he was 12 years old, Chris was given a set of keys by one of his teachers: 'If you want to use the telescope, help yourself.' It was the last day of term at Torquay Boys' Grammar School, where, a few years earlier, a couple of teachers had 'put themselves through the hell of organising teenage discos in the school hall' so that the boys could see the stars. The result of all these discos and other fundraising activities was a half-metre reflecting telescope housed in a dome that rotated and opened: an exceptional facility for a school. And that summer, Chris and a few friends moved in – a small group of 12-year-old boys, camping out at the school observatory, hoping to discover something new.

'It helped that pizza could be delivered to the observatory,' Chris told Jim. 'We discovered that early on.' Before long they were nocturnal, waking up to type instructions into a BBC Micro and watch the telescope swing into action. It didn't always work. The computer couldn't cope with the number four and, as a consequence of this quirk, anywhere in the universe that had the number four as one of its co-ordinates was out of bounds.

'Apart from that, it was a fabulous facility,' Chris said. 'I wasn't trying to take pretty pictures of the night sky. I wanted to do something real. We didn't get very far. But we tried to use this thing to tell us something about the universe that we didn't know.'

He had joined the school astronomy club mainly so that he could get a pass to jump the queue for lunch, but a very friendly talk by the astronomer and veteran broadcaster Patrick Moore had sucked him in. He remembered sitting 'in terrified silence', listening to Patrick talk about how little was known about the outer planets. 'At this point this idea went *zing* in my brain,' Chris said. 'I suddenly realised that science wasn't about reciting things from textbooks.' It was an ongoing pursuit.

'It wasn't all cut and dried,' said Jim.

'No. Exactly!'

An adult standing on a stage telling him what we didn't know was much more exciting to Chris than being told what was already known.. Soon afterwards, he wrote to Patrick: 'Dear Patrick, I am 11, we have just met . . .' and Patrick responded on a little postcard. And the correspondence continued happily for many years. One postcard simply said, 'Chris. Yes. Haste, Patrick.' By the time it arrived, Chris couldn't remember what the question was. But that wasn't the point. It was thrilling to be corresponding with Patrick Moore, star of the BBC TV show *The Sky at Night*! Another little postcard arrived addressed to 'Chris Damn I Can't Remember Your Surname Esq.', which he thought was 'rather nice'.

In Sixth Form, he spent a summer at Bayfordbury Observatory in Hatfield, part of the University of Hertfordshire. 'A bit of money' from a Nuffield Science Bursary helped to support him while he was there. He was just running computer programs, 'but it didn't feel mundane', and at lunchtime he got to talk to research scientists in the canteen. No one in his family had been to university, so it was helpful to be able to find out what higher education, and scientific research in particular, was all about. 'It's not as if I had parents who were doing this sort of thing,' he said.

He won a $500 Earth and Space Sciences Award when he was still at school for an article he wrote on 'Cosmic

Dust Around Young Stellar Objects' and was sent off to Philadelphia 'to go and compete with the Americans'. 'Meeting other 17-year-olds who were all excited about doing actual science' helped him to realise that he was 'on the right track'. And, his confidence suitably boosted, he applied to Cambridge to read natural sciences.

He chose geology as one of his options but was 'fairly useless' at it. 'I remember one exam when you had to identify ten rock samples in half an hour and I resorted to tasting them,' he said. One of the samples was rock salt, so it was not entirely absurd. 'But I got a note in the examiner's report pointing out that these samples had been in use for 80 years and probably candidates should be discouraged from licking them.'

'That's my contribution to geology,' Chris said. 'And then I became a physicist ... These days astronomy is mostly astrophysics, so a theoretical physics background was a good idea. Also, through watching shows like *Horizon*, I was fascinated by quantum theory and general relativity and wanted some of that.'

Stars were more entertaining than rocks and he started 'popping down to Patrick's house to use his telescopes'. Patrick later introduced him to the Queen guitarist Brian May, and the three of them wrote a book together, *Bang! A Complete History of the Universe*, when Chris was meant to be doing his PhD. Chris and Brian would drop everything to attend writing weekends arranged by Patrick, only to find themselves obliged to look for Patrick's much-loved cats. (Ptolemy, in particular, was always disappearing.) Apparently 'nothing else could be done unless the position of the cats was precisely known'. The other barrier to progress was the cricket. 'So we watched The Ashes and looked for the cats, as the publishers tore their hair out.' Eventually, they sat around a table going through Patrick's 80,000-word splurge, sentence by sentence. 'So the voice in that book doesn't sound like Patrick or me or Brian. I couldn't tell you which sentence was written by who. It was great fun to do. It was like being in a band with Brian May.' Rock-and-roll lifestyle notwithstanding, Brian had 'never experienced anything quite like it'.

There are at least 100 billion galaxies in our universe: swirling cosmic collections of dust, gas and billions of stars held together by gravity. The Milky Way has 100 billion stars. Galaxy M87, one of the most massive galaxies in the near universe, contains 10 trillion. A galaxy with just a billion stars is considered small. And the question that preoccupied Chris when he graduated with a degree in theoretical physics was this: how do stars form?

'The two-line description fits perfectly,' Chris said. 'A cloud of dust and gas collapses under its own gravity and eventually reaches the density at which nuclear reactions can start. That's how a star forms.' Cosmic dust coalesces, a hot core forms and nuclei fuse – and out of cold, dark space bright stars are born.

'But the three-page version doesn't work.' Why do particular stars form where they do? What triggers star formation? Why do some stars shine brightly and others fail? We don't even know for sure which came first: the galaxies or the stars.

Soon after he started his PhD in Oxford, Chris decided that his degree in theoretical physics would only get him so far if he wanted to 'untangle the whole history of these newly forming stars'. It was a messy problem and measurements of density, temperature and pressure seemed superficial. 'As a physicist you're rather limited,' he said. 'All you can really see is this blob inside which you know a star is forming.' To understand what was going on inside the stars, he needed to study their chemistry, to find out how 'a hundred or so chemicals' were reacting.

Nuclear reactions make stars shine. Energy is released as light when, for example, two hydrogen nuclei (atomic number 1) fuse to create helium (atomic number 2). Or a hydrogen atom and a helium atom form lithium (atomic number 3). This energy also prevents shrinking clouds of cosmic dust from collapsing further. These are the reactions that make stars emerge and shine for a while (anything from a few million to 10 billion years) before they go supernova, becoming a 'white dwarf', or disappear into a black hole.

And so Chris decided to study astrochemistry for his PhD but, after six months of listening to many fascinating talks about galaxies, he realised 'with mounting horror' that he hadn't done any original research. 'And no prospect of doing much was on the horizon.'

There was no shortage of data. The Sloan Digital Skies Survey (SDSS) had just released nearly a million high-quality images of galaxies that had been spotted by a robotic telescope at Apache Point in New Mexico. 'For the first time we could do statistics with the galaxy population,' Chris said. He wanted to use this big data set to study the relationship between galaxies and stars. 'But the first thing we needed to do before we could worry about star formation was to think about shape. The shape of a galaxy tells you about its history.' It reveals mergers with

> *'The shape of a galaxy tells you about its history'*

other galaxies, often violent, and acquisitions of stellar subsidiaries. 'It tells you when and where the galaxy has pulled material in from its surroundings. It may even tell you about the history of star formation in that galaxy. So, shape is fundamental.'

The stage was set. Chris had access to powerful supercomputers to help him process the high-resolution images from SDSS. 'I could sort them out,' he thought. Put elliptical galaxies in one group, spiral galaxies in another. 'Then perhaps I could start to ask some questions: OK, these are the spiral galaxies. They all live in the same sort of place. They all have the same number of spirals. They have the same kind of environment. Do they have the same kind of star formation? And these are the elliptical galaxies, big balls of stars.'

Chris then interrupted himself to confess: 'I'm making all this sound like there was a plan. And there really, really wasn't.'

Understanding how stars form was 'a beautifully messy problem' – attractive but difficult to pin down. Still, an obvious first step was to sort the galaxies according to their shape: elliptical, spiral, merger or dwarf. 'We needed to take our million galaxies and be able to split them up into these little groups.' Belatedly he set to work, only to find that step one of sorting out the universe – the apparently simple task of classifying galaxies according to their shape – was hard, 'unexpectedly hard'.

A child can spot the difference between an elliptical and a spiral galaxy. Computers struggled. Security checks that ask us to recognise a set of wonky and distorted letters to prove we are not robots exploit this digital weakness. 'Computers are terrible at this,' Chris said. And it would take him too long to do it himself. The literature was strewn with the bodies of PhD students who had tried and failed.

'If you have 20 galaxies to classify, a professor can do it. If you have 2,000, you can get a student to do it. If you have a million, you have a problem pretty quickly.'

Kevin Schawinski, another Oxford PhD student at the same time as Chris, was looking for elliptical galaxies that were blue.[1] Clicking through images of galaxies like a man possessed, he had classified a galaxy every four seconds for eight hours solid on seven consecutive days, before collapsing in an exhausted heap with just 5 per cent of the job complete. Chris contacted Kevin, having heard about his heroic attempt at galactic shape sorting, and they met for a pint in the Royal Oak to discuss the problem. Even if Kevin could keep up his frightening pace non-stop for five months and get the job done, there were better ways of using his brain. If only computers were a bit smarter or people were as plentiful as Pentium chips.

The solution, when it came a couple of pints later, was, in retrospect, obvious. If they didn't have time to do the work themselves and computers were incompetent at this task, the only way to get the job done was to involve as many people as possible and get them to help. NASAstardust@home was doing something similar, inviting people to look for evidence of stellar dust on highly magnified images of a sticky surface on the Stardust Interstellar Dust Collector. 'I remember thinking: if people will look at grains of dust, then surely somebody will help us look at these galaxies,' Chris said.

1 Typically, elliptical galaxies are red and spiral galaxies are blue, but Kevin had found plenty of exceptions and thought 'the elliptical blues' might hold the key to understanding how galaxies transitioned from spirals to ellipses.

In 2007, Chris and Kevin created Galaxy Zoo. They made all the images from the Sloan Digital Skies Survey available online and launched a website that invited visitors to see galaxies that had never been seen before. Until a human clicked on the relevant icon, hundreds of thousands of data files from the SDSS server remained unopened and those spectacular images of far-flung galaxies had not yet been seen by humans.

The response was immediate and overwhelming. Within 24 hours of the website opening, galaxies were being classified at a rate that was 70,000 times faster than what Kevin had achieved working like a madman, solo. An early surge in demand caused the SDSS server to crash. Fortunately, someone had made a copy and the back-up server saved the day. Twenty million classifications of galaxies were made in the first six weeks.

The process was simple: 'Look at this galaxy, click a button and tell us what shape it is.' Visitors would choose a galaxy and then be asked a series of questions: Can you see spiral arms in this galaxy? Is there a bulge at its centre? Is there anything odd about this galaxy? Their answers could then be analysed and cross-referenced with observations made by others. People do make mistakes (even professors), but if five Galaxy Zoo visitors agreed about the shape of a galaxy, Chris and Kevin decided the data was reliable. Later they checked for systematic bias and found a marked tendency to decide that spirals were right-handed when in fact they spiralled to the left.[2]

'Collectively our volunteers are actually better than Kevin at doing this,' Chris said. 'They're better than

2 Perhaps because right-handers, who are in the majority, subconsciously want the universe to go their way.

professional astronomers and actually better than computers . . . You don't need to know what a galaxy is. You don't need to be an astronomer. In fact, there's some evidence that it helps if you're not. It's this process of collaboration that we gradually started calling Citizen Science.'

'You don't need to know what a galaxy is. You don't need to be an astronomer. In fact, there's some evidence that it helps if you're not'

'In some ways this approach of yours makes the full democratisation of science possible,' Jim said. 'It's almost a social movement. Is that putting it too strongly?'

'I think that's the aim,' Chris replied. 'Whether we're there yet or not, I'm not sure. I want people who've never thought about science to have a moment with Galaxy Zoo, or one of the other projects, when they suddenly realise that what they're doing is valuable and real.' Nothing motivates Chris more than the people who, when he meets them down the pub, tell him they hated science at school. And having encouraged these people to visit Galaxy Zoo, he hopes they will get hooked.

'What do we now know about galaxy formation that we weren't sure about before you started?' Jim asked.

'Lots of small things, I think, is probably the best way to put it,' Chris said. 'It's true that most star-forming galaxies are spiral. But, actually, there are lots of star-forming ellipticals as well. Equally, most big red galaxies are elliptical, but there are lots of red spirals too.' In short, Chris and the citizen scientists have shown that galaxy formation is much more complicated than was previously thought. A conclusion like this can sound a little disappointing, but not to Chris: 'It's a great result for an observational scientist. That's exactly what I want to hear!'

'And [Citizen Science] doesn't stop with extra-galactic astronomy,' said Jim. Galaxy Zoo has now become part of a wider Citizen Science initiative, the Zooniverse. Some Zooites study galaxies or hunt for planets. Others participate in Snapshot Serengeti, classify bat calls or watch plankton.

'We have a project looking at classifying plankton, and for the first time we've seen one plankton eating another type of plankton. We didn't know those two types of plankton were predator and prey. And now the world does because of one citizen scientist's discovery.'

'The most reassuring thing is when you ask our volunteers why they do this,' Chris said. 'When you ask them: "Why do you spend your time classifying galaxies when you could be doing goodness knows what on Facebook?", they say that they want to make a contribution. There is something pleasing and satisfying about doing something that is real.' It's the reason Chris became excited by science during that summer he spent in the Torquay Boys' Grammar School observatory, aged 12. 'It changes everything for me,' he said. 'I'm happy to stand up on a stage and talk to the audience about the wonders of astrochemistry or the beautiful research I did last time I went to Hawaii to use their telescope, but it's much more fun to stand on a stage and talk to the audience about what people like them have done.'

In an age of digital data and the internet, ivory towers and laboratories have their uses but they are not essential. Anyone who is curious can be a scientific hero. Chris wants his citizen scientists to be empowered. He wants them to say, 'Hang on, there's something weird in this image.' Hanny Arkel, a teacher in the Netherlands, is a fine example. 'What's the blue stuff below the galaxy?' she wondered, and it led to the discovery of a new phenomenon – a quasar ionisation echo named Hanny's Voorwerp, in her honour.

Members of the Planet Hunters (a sub-group within Galaxy Zoo) take things one step further. They are working at graduate level, finding planets around other stars. 'We should probably think about giving them a degree for what they're doing,' Chris said. 'They're certainly publishing papers.'

Galaxy Zoo offers anyone who's interested the chance to experience the highs and lows of doing research. 'We get the eureka serendipitous discoveries and the slow steady build of knowledge that gives people a sense of what real science is like.'

'When it comes to astronomy, would you say we're living in a golden age?' Jim asked.

'It's a wonderful time to be interested in this stuff,' Chris said. 'Maybe we're not quite up there with Galileo and the invention of the telescope. But we're discovering planets around other stars. We're discovering what happened in the first trillionth of a second after the Big Bang. Astronomy is very fashionable. We are able to share these fabulous stories with a large number of people, and the key thing is to get them to go on, not just to read a newspaper or listen to a radio programme, but to participate in those discoveries as well.' And looking ahead, Chris thinks astronomy is going to need all the help it can get. The Sloan Digital Skies Survey is just the beginning. Several more digital surveys are about to come on stream.

'There are surveys coming up in astronomy that will produce not just so much data but so many weird things that the capacity of professional astronomers to follow up on weird things will be exhausted,' Chris said. 'We are going to need a large community of citizen scientists, of amateur astronomers using robotic telescopes, for example, to tell us what's worth looking at. It's great fun to think about how to build that community.'

Citizen scientists will be needed to analyse oddities as they emerge, if we are to be able to follow up all the interesting things we will find in the universe and not get overwhelmed. And the more challenges the citizen scientists take on, the more they will be empowered.

'I take the "Citizen" in Citizen Science very seriously,' Chris said.

'How many people have taken part in the research now?' Jim asked.

'We just topped a million users,' Chris said. 'So in seven years, that's pretty good going.' In 2013, people sitting at home accounted for about 300 years' worth of round-the-clock effort in the research projects in the Zooniverse. 'It's an opening up of the scientific process that hasn't happened before. And in my grander moments, I think this will have lasting consequences.'

The teachers at Torquay Boys' Grammar School gave Chris the opportunity to do some scientific research of his own. Now he hopes Citizen Science projects will give everyone, not just the well-facilitated few, the chance to do the same.

'If I were at school now, I wouldn't need to be at a great school with a large telescope,' he said. 'If we can keep that as a sustainable part of how science is done, I think that will be a very exciting thing.'

Courtesy of Jane Francis

JANE FRANCIS

'There are lots of women in Antarctica today'

Grew up in: Kent and Wiltshire
Home life: sister and two brothers, and a large family of cousins, nieces, nephews
Occupation: paleoclimatologist
Job title: Professor of Palaeoclimatology and Director of the British Antarctic Survey
Work address: the British Antarctic Survey, Cambridge
Inspiration: a tiny fossil leaf
Passion: Antarctica
Mission: to understand the global climate, past and present
Best moment: 'A perfect day in Antarctica with no wind, blue skies, bright sun and twinkling icebergs'
Advice to young scientists: 'Say YES to new opportunities'
Date of broadcast: 21 March 2015

Antarctica is Jane Francis's expertise and her addiction, but her research career began in a gentler climate, in Dorset studying the rocks on the Jurassic Coast. A chance encounter with a tiny fossilised leaf pushed her into the study of long-dead plants, and her intimate knowledge of fossilised leaves and petrified trees has taken her all over the world, studying ancient climates by pulling together threads of evidence from rock, minerals and fossilised plants. She caught Antarctic fever on a geology expedition in 1989, and in 2002 she was awarded a Polar Medal for her contribution to polar science, joining past winners Captain Scott and Ernest Shackleton. In 2013 she became the Director of the British Antarctic Survey, part of the National Environment Research Council, and is responsible for five ships, two aircraft and five polar research stations, as well as hundreds of British scientists who are working on this icy landmass the size of Europe.

Pitching a tent on ice isn't everyone's idea of fun, but Jane Francis likes nothing better than camping in Antarctica for a couple of months.

'What is it like living in minus 20 degrees, or often much colder?' Jim asked.

'Actually, I like being cold,' Jane replied. 'It's really a dry cold, that's the thing. I don't like humidity, so Antarctica is perfect for me.' She has spent the equivalent of two years of her life living in a tent in the polar regions.

Jim wondered if there were times when being the only woman in the team was awkward. 'Is it ever difficult for you, living in the cramped confines of a tent with male colleagues?' he asked.

'No. I've never found it to be an issue myself,' Jane said cheerfully.

Jane studied geology at Southampton University, then looked at jobs in the oil industry. It was a well-trodden path for geology graduates, so the small print in one of the job descriptions came as quite a shock. 'I distinctly remember this. It said: women need not apply. It was a bit depressing.'

Job offers from multinational companies 'didn't feel quite right' but the chance to do a PhD at Southampton University did. Her subject was the coastal rocks of Dorset between the isles of Purbeck and Portland, where the remains of Jurassic forests are amazingly well preserved. (The coastline is now a World Heritage Site, but it wasn't then.) Stumps of petrified trees look eerily modern, as if they died just a few years ago, not many tens of millions. Dinosaur footprints are still visible and in disused quarries on the Isle of Portland, layer upon layer of sedimentary rocks are beautifully exposed.

Jane could travel forwards and backwards in time just by glancing up and down. 'It was like reading the pages of a book, layer by layer, as you go up through the quarry section,' she said. She spent many days sitting in the quarry, splitting rocks to examine the sedimentary composition, hoping to gain more insight into a time 150 million years ago when the sediments that are now in Dorset were just 35 degrees north (about the same latitude as Morocco today).

Another day, another rock. But one day she spotted tiny fossilised leaves as she split open a fragment of a boulder. 'I remember sitting there thinking I could either throw this over my shoulder into the sea and carry on with my thesis the way it was supposed to go, or I could open up this rock again and I could work on these leaves and it would take me in a completely different direction.'

'I remember sitting there thinking I could either throw this over my shoulder into the sea and carry on with my thesis the way it was supposed to go, or I could open up this rock again and I could work on these leaves and it would take me in a completely different direction'

She was a geologist, not a botanist. She studied rocks not plants. But finding fossil leaves in Dorset's Jurassic forest was a very rare treat, the kind of evidence that no self-respecting scientist could ignore. A few minutes later she looked at the fossils again, took a deep breath and her career as a palaeobotanist began.

Once she knew what she was looking for, Jane went on to find many more of these tiny fossilised leaves. Fossilised cones of monkey puzzle trees followed. She even found evidence of ancient pollen, dating back to the days of the dinosaurs. When she examined thin slices of petrified wood under a polarising microscope, ancient tree trunks came alive. They looked as if they had just been felled and had been fossilised so fast that even delicate structures remained intact. She could see the cell walls and noticed that growth rings were narrow, suggesting that conditions were not ideal for growing trees. The rings also varied greatly with the seasons.

Slowly, the precise nature of these trees emerged. Previously the co-existence of a coniferous forest next to warm, salty lagoons had created some confusion. What sort of climate could support both of these apparently contradictory features? Jane's research helped to resolve this conundrum. The Purbeck forest in the Late Jurassic, she concluded, had a Mediterranean-style climate with warm, wet winters during which the trees, similar to modern junipers, could grow, and hot, arid summers when hyper-saline lagoons formed.[1]

Her PhD thesis on 'The fossil forests of the basal Purbeck Formation (Upper Jurassic) of Dorset, southern England' was completed in 1982. Invitations to study fossil forests

---

1 Quite wonderfully, I thought, when Jane arrived at the BBC building in W1A, she pointed out fossils in the Portland stone, and showed just how full of life a block of rock could be. The marine life in the Late Jurassic is preserved in the stone walls of New Broadcasting House: a profusion of scallops, oysters and snails have all left an impression in the rock that was formed in warm, clear waters 150 million years ago.

in Australia, the Arctic and Antarctica followed. She had 'a fantastic time' working in Australia for several years, pulling together different threads of evidence from rocks and fossilised plants and trees to describe ancient climates. Apparently 'Australia was quite cool a hundred million years ago.'

The Canadian High Arctic was next. A forest of mummified trees had been discovered and Jane's expertise was needed. As she walked over piles of 50 million-year-old leaves, they crunched underfoot, 'crisp like a deciduous forest. The forest looked like it had been buried yesterday. It was amazing preservation.'

'So, for 50 million years no one had set foot in this forest?' Jim said, impressed.

'We were the first people,' Jane said, then qualified her enthusiasm a bit. A couple of geologists had found the fossil tree stumps the year before but, knowing nothing about ancient forests, had not stayed long. And so a group of scientists were invited to start working on them.

'I bet you felt pleased that you hadn't thrown that rock with the fossilised leaf off the cliff edge?' Jim said, smiling.

'Absolutely!' Jane exclaimed. 'That certainly opened some doors for me!'

Jane made her maiden voyage to Antarctica in 1989, on a geology expedition run by the British Antarctic Survey.

'Was it love at first sight?' Jim asked.

'It was indeed. I got Antarctic fever from then on and can't keep away!'

On clear days, the sight of icebergs set against intense blue skies would send her spirits soaring. The research opportunities were fantastic, but when the weather was bad, precious research days disappeared and were spent zipped up in her sleeping bag, waiting for blizzards to pass. All observations were put on hold.

There's plenty of space on the continent but solitude is elusive. Team members work together all day for safety, and then sharing a tent means that there is little time to oneself.

'You know there's going to be someone in the team who gets up your nose,' Jane's friend and colleague, Joe Cann said. 'You just hope there won't be more than one or two.' Jane, however, seemed wonderfully collegiate and sunny. It was hard to imagine her losing her temper or sulking.

'What do you do if you're cross with a colleague or something they've done?' Jim asked.

'I tend to let people know I'm going in my tent. "I'm going into my sleeping bag, please don't disturb me!"' Jane replied. 'And I stay there for a few hours, listening to music or reading a book.'

She spends a lot of time thinking about food. Dinner is a major event: normally rehydrated beef granules with instant mashed potato and plenty of chilli. There's no shortage of chocolate and she eats vast quantities of it throughout the day to provide the energy needed to keep warm. But she craves apples and lettuce. Anything with a crunch. 'I started dreaming about celery,' she said.

Pressed by Jim to share some of the less glamorous aspects of being an Antarctic scientist, Jane said, 'Not being able to wash my hair was the thing I probably hated most. But otherwise I really loved it.' She didn't mind wearing the same clothes 24/7, day after day, even if it did 'get a bit uncomfortable after a while'. 'There's no point in washing, actually,' she said. 'When you're cold you don't smell too bad!'

'Thinking about it,' Jim reflected, 'Antarctica must be the hardest place on Earth to study rocks and fossils. So, if you'll forgive me for putting it like this, why do you bother?'

'Antarctica is an amazing place,' Jane said. It is more than 99 per cent ice but there are some rocks exposed, on the tops of mountains sticking up above the ice cap and around the coast. And these rocks date back to a time when Antarctica wasn't covered in ice. A hundred million years ago the South Pole was warm and humid. There were tropical rainforests in Antarctica and Amazonian-style vegetation: 'It was quite a different world.'

'If I took you there,' Jane said, 'you can bet your bottom dollar that you would find a fossil leaf very quickly.'

'Is it true that penguins evolved when there were trees on Antarctica?' Jim asked

'Yes, that's a really great fact,' Jane said enthusiastically. 'I like telling people that!'

Buried in sediments that are 50 million years old, Jane and her team found big tree stumps and lots of fossil leaves. They also found piles of penguin bones, belonging to many different species. All sorts of different penguins thrived in Antarctica before the ice sheets set in, waddling through sub-tropical forests, not jumping like lemmings into freezing seas. 'As the climate subsequently cooled, most of the species moved north to South Africa and the Galapagos Islands, for example. Only a few species, mainly Emperors, stayed in Antarctica.'

'Presumably they liked it there?' Jim suggested. 'They had the chance to emigrate . . .'

'Either they liked it or they were stupid!'

In ways that Jane could never have predicted when she picked up those fossil leaves in Dorset, palaeobotany is now being used to help predict future climate change, as we confront the consequences of all the carbon dioxide and other greenhouse gases that we have pumped into Earth's atmosphere and strive to understand what might happen next. We have created some impressive mathematical models of climate change, based on vast quantities of data and using some of the fastest supercomputers[2] in the world. But these models do need to be tested, and as there is no data available from the future, one solution is to get computer models to simulate conditions in the past. The results of these 'hindcasts' can then be tested against data gathered on the ground by paleo-climatologists such as Jane, and the ability of the model in question to make accurate forecasts can be assessed. The study of ancient climates has never felt more cutting-edge.

'We've also got to go out into the field and make more observations,' Jane said, 'to test the models to see if they are right or whether they need to be changed in some way to really simulate Earth's systems properly.' Satellites have an important role to play, gathering data on ice sheets from space. Autonomous vehicles are doing a great job, especially in our oceans. But nothing beats people visiting in person.

What happens, Jim wondered, when Jane tells the computer scientists that their predictions don't match the geological evidence? 'If they've spent years developing these models, I can imagine they wouldn't be too happy . . .'

2 One of the newest climate computers can do a quadrillion calculations every second.

'Sometimes they tell us that our geological evidence is wrong,' Jane said, 'or that we've interpreted it incorrectly.' Collaborating was never going to be easy between scientists with such different skill sets and interests. 'But we work together more these days,' Jane said. And it's a powerful combination when they do. 'We're getting better and better at climate modelling.'

The global climate has changed dramatically over deep geological time, for reasons that have nothing to do with us. Exploding volcanoes emit vast quantities of greenhouse gas, for example. Our entire planet was probably frozen over 650 million years ago, during 'Snowball Earth', and there's evidence of other global glaciations. There have been tropical forests on the South Pole. 'And the geological record has an important story to tell. It helps us to understand how Earth's systems respond to big changes. How they responded in the past and what they can tell us about how things might happen in the future.'

Since 2016 the concentration of carbon dioxide in our atmosphere has been more than 400 parts per million. The last time this was the case was about 3 million years ago, just as our earliest ancestors (as far as we know) first walked on the African Savannah. 'The last time our planet had 400 parts per million of carbon dioxide in the atmosphere, we know from the rock record that the polar ice sheets were more changeable,' Jane said. 'They waxed and waned quite markedly. And it does suggest that we are entering a period now when we might expect the ice sheets in Antarctica to be much less stable than they are today.'

A major part of scientific exploration in Antarctica today is focused on trying to understand how the ice sheets are responding to our warming climate. And there is a lot more to be concerned about than stranded polar bears. If we want to understand how the global climate will change, then understanding what's happening in Antarctica now is pivotal. Antarctica, 'that big block of ice, that big refrigerator at the bottom of the world', exerts enormous influence on a number of global systems. 'It affects sea level. It strongly affects ocean and atmospheric circulation. It has a fundamental impact on the whole Earth system.'

'If we want to understand how the global climate will change, then understanding what's happening in Antarctica now is pivotal'

The West Antarctic Ice Sheet, in particular, is a cause for concern. Most of this ice sheet is underwater and measurements show it could be melting from below, as warm water percolates through its underside. 'If it reaches a critical point, the whole ice sheet could melt very quickly,' Jane said. 'This translates into a three- or four-metre rise in sea level globally.'

'A sea level rise of several metres would put most of England under water,' Jim said. 'Within what sort of a—?'

'That's what we're trying to find out,' Jane said before he had time to finish his question. 'Exactly what is going to happen to this ice sheet? How long is it going to be stable?'

Jim suggested that despite all the valuable work being done by climate scientists, geologists and mathematical modellers, our knowledge of the global climate is just the tip of the iceberg.

'And the iceberg is melting fast,' said Jane. 'So we've got to do more work to find out what's going on.'

Scott Barbour/Getty Images

JOHN SULSTON

'You need to be obsessive to do certain types of science. I'm not alone in that'

Grew up in: Rickmansworth
Home life: married to Daphne with two children, Ingrid and Adrian, and three grandchildren, Micah, Kira and Madeleine
Occupation: molecular biologist
Job title: Founding Director of the Wellcome Sanger Institute
Inspiration: noticing that the number of neurons in a tiny worm increased after the larva had hatched
Passion: the molecular biology of *C. elegans*, a tiny nematode worm
Mission: to sequence first the worm and then the human genome and make sure the information was available to everyone for free
Date of broadcast: 29 November 2012

John Sulston died in March 2018

John Sulston was the winner of the 2002 Nobel Prize for Physiology or Medicine for his work on the developmental biology of a tiny transparent soil worm, *C. elegans*. Staring down a microscope at this unassuming organism for months on end, he traced the origin of every one of its 959 cells by watching them divide and noticed that certain cells were programmed to die. He pioneered genomics – the study of 'the whole thing', not just individual genes – and produced an ordered list of all the base pairs, or genetic letters, on the worm's DNA double helix. (There are 100 million.) Spurred on by his success, he urged the Wellcome Trust to invest heavily in a new centre dedicated to the study of the human genome in the UK. As founding director of the Wellcome Trust Sanger Centre (1992–2000), he led Britain's involvement in one of the most exciting international research collaborations in modern science. The Human Genome Project is a feat of biological exploration that has been compared to the *Apollo* missions to the Moon, and John fought tooth and nail for our newly found knowledge of the human genome to be made available to everyone for free.

'Actually, that was the best time of my life,' John Sulston said about the years he spent staring at a tiny nematode worm named *Caenorhabditis elegans*, on account of the elegant way it moves, glued to a microscope all day, every day, for four hours at a stretch.

'I could look away for a minute, that's all,' John said, clearly delighted with the arrangement. Jim looked concerned. 'It was all right,' said John with an encouraging smile. 'I enjoyed it.'

Some scientists do get stuck worm watching without joy. The task becomes boring and repetitive. They've already spent ten years of their life doing exactly the same thing on a production line and they've got another 30 years to look forward to. 'That *is* sad,' John said definitively. 'But you don't do it mindlessly!' he exclaimed. 'That's nuttery! If you're making progress and it's for shorter periods of time, that's completely different. You get very good at a particular task, and you do it over and over again.'

'For a year and a half,' Jim noted.

'Yeah. A year and a half,' John said, as if the years were minutes. 'It's extremely satisfying. It's artisanal.'

As a young man John didn't seek responsibility. 'Is it fair to say you tried your best to avoid it?' Jim asked.

'Oh yes,' said John with disarming honesty. 'I like working on my own, always did.'

Later in life, when he was Director of the Wellcome Sanger Institute, responsible for hundreds of staff and budgets worth billions, and in charge of the Human Genome Project in Britain, he would escape, whenever he could, to study the molecular biology of *C. elegans*.

'That was what I regarded as my real job,' he said. And there were other reasons why he remained wedded to the worm. 'I think it's very useful for executive-type people to have real jobs to do,' he said. 'It kept me sane. It really did.' Worm therapy, if you will.

'I think it's very useful for executive-type people to have real jobs to do'

A chemist by training, John joined the Laboratory of Molecular Biology (LMB) in Cambridge in 1969 to work under Sydney Brenner, who ran the cell-division department with Francis Crick and was the charismatic champion of a tiny worm that he had rescued from obscurity in 1963.

Despite being less than a millimetre long, *C. elegans* has nervous, digestive and reproductive systems and, rather wonderfully for anyone interested in seeing inside them, its cells are transparent. This sophisticated see-through worm made life easy for classical geneticists too. Self-fertilising hermaphrodites that have hundreds of children are a great help to anyone studying inheritance. Even better, those children are ready to have children of their own just a few days after they are born.

Cell division was a hot topic in molecular biology in the 1970s as scientists struggled to understand the extent of DNA control. Sydney's overarching goal, communicated in his famously entertaining daily talks during the mid-morning coffee-break, was to understand how a complex multicellular organism like the worm evolved from a single fertilised cell. In particular, he wanted to know how specialised cells (in the digestive or reproductive systems, for example) know when to stop dividing. Was this a process that was controlled by DNA?

John focused on the chemistry of the worm's nervous system, trying (and generally failing) to identify the genes responsible for the creation of different neurotransmitters. After staring at the worm's nerves for many months, he made a surprising observation: he noticed that the number of neurons in the worm's ventral cord increased from 15 in newly hatched worms, to 57 in fully grown adults. Cell division was widely assumed to take place only in the eggs (the worms appeared to hatch fully formed, after all), but John's observation proved that worm cells were dividing after the eggs had hatched. How else could the number of neurons increase? And it made him think. Perhaps he could observe cells dividing in the larva? By shifting the focus from egg cells (which, frustratingly, were opaque) to transparent larva, maybe he could succeed where others seemed to be failing and so help to realise Sydney's dream?

The prospect was both exciting and stressful. To take the pressure off, he played mind games with himself. 'You pretend you're not taking it seriously,' he said. He went into the lab on a Sunday afternoon, telling himself it was just for fun. There was nothing to lose if things didn't work out. This was not the main focus of his research.

Being able to see into the cells certainly helped, but there were other problems. The young worms squirmed and squiggled and kept slipping off the glass plate under his microscope before John could focus. 'This was very much biological messing around.' And it didn't always go well. Various efforts to keep the worms in one place resulted in mass injury and death. John's trick, established after

'I wanted a happy worm so I could see the cells divide'

months of playful but determined trial and error, was to draw a very thin layer of agar jelly over the microscope slide and then add some bacteria for the worm to eat. 'I wanted a happy worm so I could see the cells divide,' he said.

He nudged these slippery organisms into place, first with sharpened toothpicks and later with platinum wires. 'This is the real way of doing things. Simple devices that make things work.' He then laid a cover slip gently on top of the worm, so that it was 'slightly confined but not squeezed'. It could still move about in its elegant, sinuous way but generally preferred to stay near the delicious agar jelly. And while the worm munched away, John was able to watch its cells dividing. Both J. Sulston and *C. elegans* were happy. 'The very first time I saw a cell divide and saw its daughters divide, I was ecstatic.'

'What does an ecstatic John Sulston look like?' Jim asked.

'He tends to go a bit quiet,' John replied. 'Then he goes out and makes a lot of noise and annoys everyone around him.'

Having seen cells divide once, John was hooked, determined to witness every cell division and death that occurred as 15 cells became 57. He persuaded Bob Horvitz, a geneticist who was visiting from MIT, to watch worm cells dividing with him. Initially Bob didn't see the point, but before long he too was addicted. Together they mapped out which cells came from where and in 1977 published the entire lineage of the larva. They also showed that the whole process was precisely timed and, in so doing, demonstrated strong evidence of DNA control.

It was ground-breaking work, but no sooner had John achieved what, many years before, he had set out to do, he wondered if it really was such a worthwhile occupation after all. He considered leaving science altogether and applied for a job with the government fisheries department in Lowestoft. 'I suppose I hankered after doing something different with my life,' he said. 'Perhaps it was a down moment, a classic mid-life crisis. But I wanted that job, I really did. I thought it would be fun.' Sydney, however, told him he was 'being ridiculous' and refused to give him a reference. John acquiesced, accepting Sydney's authority without much of a fight.

Having traced all the cells in the adult worm back to the moment of hatching, the obvious next step was to try and find out what happened in the egg. Could every cell in the young larva be traced back to a common ancestor, a single fertilised cell? Strangely, rather than jump at the chance to finish the job, John decided to wait. His colleague Bob Waterstone explained: 'There had been somebody else assigned to look at the embryo and the pattern of divisions there, and John, in his very characteristic modest way, didn't want to step on anybody's toes.' Later a group in Germany got involved and 'again John stayed out of their way'. Only when it was clear that these groups had either given up or were completely stuck did John feel happy to start his research.

Back at his microscope with another lineage to complete, all thoughts of working for the fisheries department disappeared. A year and a half of solid worm watching followed, with John sticking to a strict schedule of two microscope sessions a day, each four hours long. (There's a hollowed-out dip in the wooden floor where John wheeled his chair back and forth in the LMB, leaning in to look down his microscope and out again to draw what he had seen.)

By using tricks he'd learnt when studying the larva and inventing several new ones, John did manage to succeed where others had repeatedly failed. His paper describing the embryonic cell lineage was published in 1983, a decade after he had started experimenting on Sunday afternoons. 'It changed everything,' John said.

In 2002 he was awarded the Nobel Prize for Physiology or Medicine (together with Robert Horvitz and Sydney Brenner) for this pioneering work. Nonetheless, John insisted that this much-celebrated paper is 'rather over accoladed, I think. It's quite embarrassing, really.'

While John had been watching cells divide, the classical geneticists had been getting to grips with the inheritance of different characteristics – tail shape, for example. They knew that certain genes existed (from experiments with mutant worms). The challenge now was getting hold of those genes as actual pieces of DNA so that they could be studied further. In short, the molecular biologists needed something tangible to work with. Sometimes people got lucky, but more often it was soul-destroying. 'An awful lot of people were wasting an awful lot of time, laboriously hunting through this genome until they nailed down the one they wanted. And I thought, this is ridiculous.'

While others raced to find a particular gene (sometimes in direct competition with other teams), John took a different view: 'I thought if I, or a small number of us, go away and make a map of the genome in which all the bits of DNA are cloned and laid out in order, we can then know roughly where the gene we're looking for is . . . It's a matter of investment.' John compared making a map of a genome to building a road through a desert landscape. It would help everyone to drive to their destination quickly and easily instead of driving in, getting stuck and having to be dug out. Once again, it seems, John was thinking more about others than about himself.

Not that his colleagues thanked him for it. Rather than expressing gratitude, they griped and grumbled and criticised his selfless mission. Why waste time mapping the entire genome when so much if it was junk? Not all DNA spells out genes. To attempt to read it all was overkill, they argued. 'Lots of people thought it was not very sensible,' John said. 'The worm genome is 100 million bases, or letters, long and people correctly thought a lot of it didn't do a great deal. The trouble is, knowing which bit matters and which doesn't . . . It was a gamble to say, "I think we need to get at all the genes, not just a few."'

The research didn't need much money (at least, not to begin with) and Sydney was supportive, so John got started. Never mind what anyone else thought; he thought it would be helpful. 'There were a small number of us in biology who decided that we were going to do what came to be known as genomics, looking at the whole thing.'

'There were a small number of us in biology who decided that we were going to do what came to be known as genomics, looking at the whole thing'

By the early 1980s, a group that had started life as a handful of biologists picked by Sydney Brenner had grown into a thriving and close-knit community of worm-a-holics,[1] all enthusiastic attendees of the international *C. elegans* conferences and avid readers of *The Worm Breeders' Gazette*, a much-loved magazine that was packed with the latest nematode news and gossip and a regular *C. elegans* cartoon.

John announced his intention to map the genome of *C. elegans* to his fellow worm scientists in 1982 and they were enthused. They liked and trusted John and it created a virtuous spiral of sharing behaviour. 'It was an excellent example of how data sharing and collaboration in science propels it,' Bob said. 'John let everyone know that it was going to be an open and fair playing field.' And everyone believed him. His gentlemanly behaviour over the embryonic lineage (waiting for others to finish, not rushing to be first) had won him many friends, and instead of jealously guarding their discoveries, scientists happily volunteered data.

Other scientists, John quickly discovered, had mapped isolated, scattered patches, detailing 'little tiny bits of it', but vast areas were unknown. 'I had to do the joining up,' John said as if it were a simple job. 'That's very appealing to me.'

Seven years later his map of the worm genome was complete. 'I don't think I'm a very intellectual person,' John said with astonishing humility. 'But I certainly can, through a sort of obsession and love of completeness, make a map that other people find valuable.'

Previously geneticists had spent years searching for the particular bits of DNA that corresponded to the genes

1 Sydney managed to attract some of the brightest scientists in the UK and beyond to work on this unprepossessing worm. One LMB scientist described him as 'the most intellectually seductive person I have ever met'.

they had identified by breeding mutants. Now genes could be located and isolated in weeks, not years. A 'ridiculous bottleneck' had been removed. And it encouraged John to go one step further. Using his low-resolution map as a guide, he then decided to try and sequence the worm genome: he wanted to list all the genetic letters as they appeared on the DNA double helix of *C. elegans*, a task that, at the time, was considered too difficult for all but the very simplest organisms.

It was an ambitious undertaking. The first genome to be sequenced was a virus named MS2. It had about 3,600 genetic letters. The worm genome, as John later discovered, has 100 million. By inventing all sorts of clever new techniques, he started to make progress. He might have started to relax when sequencing the worm genome seemed to be going well. Instead he started to think about the human genome. *C. elegans* was only ever meant to be a model organism, after all.

Tackling the human genome would not be easy. (We know now that it is 3 billion genetic letters long.) Nor would it be cheap. But it was absolutely clear to John that it needed to be done. If the sequence of genetic letters on a typical human genome was described and made available to researchers, the study of human genetics could be revolutionised, just as worm genetics had been, and our understanding of ourselves would, he hoped, be transformed. John approached the Wellcome Trust (who had funded his worm work) for funding to set up a project to sequence the human genome in the UK. It would, he said, require a new laboratory to be built in Cambridge and a large team of scientists working together towards this common goal and collaborating with colleagues in the USA. 'We were going for broke because we thought it was so urgent to press on,' he said.

Initially the omens were good. John and Bob were invited to a big meeting in London one Monday morning to defend their proposal. But on the Saturday night before the meeting, John was cycling home from work and got hit by a car. His hip was broken and he was unconscious for a while, but he woke up in hospital on Sunday morning determined to make it to the Monday meeting. He managed somehow to get himself onto crutches and John convinced the doctors that he could make the trip to London, using the fact that Bob was a doctor to persuade them, even though Bob hadn't seen a patient for 25 years. 'We got in the car' Bob said. 'John climbed in the back. I had no idea how much pain he was in, and we went down to London.' Later that day, 'he made an absolutely brilliant presentation' from a wheelchair, 'and convinced the Wellcome Trust that this was a good thing to do. These were really dire personal circumstances for him. It's just amazing that he had the fortitude and tenacity just to go ahead.'

The Wellcome Trust Sanger Centre in Cambridge opened in 1992, set up as John had imagined. Despite being the driving force behind its creation, he was 'astonished to be invited' to be the founding director. Many of the techniques invented to sequence the worm genome also worked for the human genome, as John had hoped they would. 'It's all DNA, after all,' he explained. The technology they used was not identical but it was closely related. But some of the challenges were unique to humans.

'We were getting stuck on one particularly nasty bit of DNA,' John said. He was tempted to leave a gap but most of his team were mildly panicked by the prospect.

'No, no, we can't leave gaps!' they said. 'Which was right, of course,' John said. After 'a little argument', John took charge, saying, 'All right, I'm going home. I'll find a way of making it work.'

At home he paced around. He went to the library. He started reading. Next day, as he had said he would, he came back to the team with a solution: 'I hit on a technique which is, I suppose, rather typical of me, not clever in the least, but it worked.' The plan was to smash the troublesome section of DNA into even smaller pieces by blasting them with as much ultrasound as possible, and then to re-clone these tiny pieces. It was simple and effective and it arrived just in time. 'We were really scaling up internationally to sequence the human genome and nobody had solved this problem at that point.'

'I hit on a technique which is, I suppose, rather typical of me, not clever in the least, but it worked'

John's 'not clever in the least' technique was a game-changer. It quickly became routine. If his day-to-day management style was *laissez-faire*,[2] technically he blazed a trail. And his strategy was clear. One of the first things he did was to set up a human genome consortium modelled on the *C. elegans* consortium: 'Scientifically and psychologically and in terms of understanding, it's much better to bring everyone into the same tent,' he said.

At the international Human Genome Organisation meeting in Bermuda in 1996, he made it clear that all the data from the human genome would be made publicly

2 John acknowledged apologetically that '[his] confederates had to do a lot of the real managerial work'.

available. He invited everyone to collaborate, just as he had done with the worm, but was disappointed to discover that the scientists who worked on human data were much less willing to share. Some wanted to patent the genes once they had been sequenced. 'They wanted to get rich from their discoveries,' John said, dismissive of their greed.

'Looking back,' Jim asked, 'do you think you were naïve to think [the human genome project] would work in the same trusting way as the worm research had done in the 1980s?'

'I wasn't naïve, because it did work,' John replied, irritated by the suggestion. 'My natural inclination when you have this competitive situation is to collaborate.' It makes sense to pool resources. If people want to be part of something, then surely they are enriched by working with others, 'rather than competing in some aggressive manner'.

'It's not easy to establish a consortium,' he admitted, but the idea that the genetic knowledge should be shared was non-negotiable, a core belief. 'There is just one basic reference human genome. Why compete over it? It's crazy . . .'

For several years the consortium made steady progress, but everything John stood for was shaken to the core by a shock announcement in 1998. The American biologist and businessman Craig Venter announced that his company, Celera Genomics, could sequence the human genome faster, cheaper and better than the multi-billion-pound publicly funded programme which employed thousands of scientists on both sides of the Atlantic.

Jim played John a clip of Craig Venter talking about Celera Genomics and accusing the scientists who worked on the Human Genome Project of being self-serving. 'I think the scientists involved did a lot of manoeuvring to get into positions to have those multi-billion-dollar budgets,' Craig Venter said. 'Most viewed it as an entitlement and planned to do this work over a ten- or fifteen-year period. Celera's model was to complete genome sequence and get it out there as quickly as possible. When we finished sequencing the human genome, after nine months, we published a paper in [the journal] *Science* and the data was made available to the scientific community, so it was not about keeping the data secret. Those were mantras that were created, particularly in the UK, because with this massive amount of money, they weren't able scientifically to compete with the new tools and approaches that we had.'

Twenty years on, Craig Venter's claims still made John's blood boil. 'What they had was not a lot cheaper, not a lot faster than what we were doing,' he said, clearly angry. 'And it certainly wasn't as accurate in the end. The thing is just hugely exaggerated.'

Then, sounding suddenly tired, bored of repeating arguments that he'd had so many times in the past when Craig Venter would challenge the Human Genome Project and he would be obliged to respond, he said: 'There's only one thing that matters and that's the issue of data release. He claims that the data was released to *Science*. Yes, in the end. But only after two years, not nine months . . . A DVD of the data was deposited with *Science*. It was only a draft sequence, not a complete one. The real sequence . . . is the sequence which remains on the computers in the public databases of the world.' This is the sequence 'that people have gone on working on'.

Scientists working on the Human Genome Project

shared everything just as soon as discoveries had been made. They published their findings weekly, without fail, on public databases that were accessible from anywhere in the world. A draft sequence (inaccurate and full of gaps) released on a single DVD was hardly the same thing.

'Wasn't there an issue with Venter using the data that had been made publicly available?' Jim asked.

'He was entitled to,' John replied.

'And using it to fill in his gaps?'

'He was entitled to do that,' John repeated, unprovoked. 'Venter says that our data corrupted his data and it would have been better if they hadn't used it, but they did use public data, and they did this because it was indeed a race. There was a serious race and it was about ownership. It was about ownership in two senses. One is that they wanted to own the database so that everybody would have to buy a subscription to use it. And they wanted to build a repository of biological information that would be instead of, not as well as, the public repository. The business model was to hold on to the data and to rent it out. This is why Celera's share price rose very, very high at that time. They had a stock offering which, I think, raised a billion dollars for them. People don't invest in a company like that in order to give the data away, do they?'

The other issue that worried John – 'no, more than worried' him – was that many, many patents were being filed on genes.

'Of course, they would argue that they needed to sell their product to get funding for future research . . .' Jim chipped in.

'Yes, absolutely right, and so you have to choose an appropriate way of raising money for the task at hand. As with a mousetrap, you don't patent the idea of a mouse-trap. You put a patent on a particular way of trapping a mouse. In other words, the kit, the material, the stuff – not the ideas.

'The human genome, and all the other genomes, are the basic information of biology, of medicine. That information should be publicly owned. I think this is an absolute principle. The idea of locking them up and renting them out piecemeal to subscribers – who, by the way, if you have such a business plan, are forbidden to communicate the data to one another – does that make any kind of scientific sense? No. It failed, mercifully. And Celera went down as a result.'

'It was a close call, wasn't it?'

'It was indeed,' John said. 'The race was real and, by the way, many people in the different industries affected by this – naturally, that's all the industries in biology – have said to me, "Thank heavens you won!"'

In June 2000, the progress made by the Human Genome Project was celebrated on both sides of the Atlantic. The British Prime Minister, Tony Blair, said: 'Every so often in the history of human endeavour there comes a breakthrough that takes mankind across the frontier and into a new era.' Many described it as the biological equivalent of the *Apollo* missions to the Moon. In February 2001, the results of the Human Genome Project were published in the journal *Nature*. The sequence of the human genome was 90 per cent complete. By April 2003, the job was done: the DNA code that makes us who we are had been written out in full and was 99.99 per cent accurate. A copy of our entire genome can be found inside the nucleus of every cell in our body. It is 3 billion genetic letters long. Reading out every letter in our genome, at a normal rate non-stop for 24 hours a day, would take a century.

'You've devoted a lot of your life's work to reading and understanding the manual of life, first in the worm and then in man. This is something that really defines your career,' Jim said.

'I'd slightly draw back from that,' John said, humble as ever. 'I'm a chemist, originally, and I thought if I just read the stuff, there'll be lots of people who will then understand it. People who are much cleverer than me. And that's exactly the way it has worked out. I'm the reader, others do the understanding, and that's why it's so important to release the data.'

'Are you proud of what you achieved with the Human Genome Project?' Jim asked, keen to encourage John to celebrate his success.

'I'm certainly very relieved,' John said. 'I suppose "proud" is too strong a word for it . . . I found it easier to be proud of the little things, because you can say, "They're mine" . . . I'm proud of the international Human Genome Consortium, because together we pulled it off.'

COLIN PILLINGER

'I made it happen. Bureaucracy hasn't achieved it.'

Grew up in: Bristol
Home life: married to Judith with two children, Shusanah and Nicolas
Occupation: planetary scientist
Job title: Professor of Planetary Sciences at the Open University
Inspiration: a lecture by the pioneering British chemist, John Beynon
Passion: his Mars lander, *Beagle 2*
Mission: to look for evidence of ancient life on Mars
Advice to young scientists: 'There's no such word as can't'
Date of broadcast: 27 December 2011

Colin Pillinger died in May 2014

Colin Pillinger was the driving force behind *Beagle 2*, a robot destined for Mars with a highly sophisticated instrument on board to analyse the chemical composition of Martian rocks. For years, he fought with the European Space Agency to be allowed to do what he wanted to do, inspired by John F. Kennedy's approach to the Apollo mission in the 1960s: 'We choose to go to the Moon . . . and do the other things, not because they are easy, but because they are hard.' He was infuriated by ESA's bureaucratic attitude and lack of ambition. But, by his own admission, he did his best work when he was angry, so perhaps it spurred him on. With the help of his wife Judith, a 'PR genius', dozens of British scientists and engineers, the artist Damien Hirst and the pop group Blur, he 'made it happen' and enthused the nation. When, on Christmas Day 2003, his robot lander *Beagle 2* descended towards the surface of Mars, thousands of people were glued to their TV screens awaiting its arrival.

From the outset, Colin's plans for *Beagle* 2 were ambitious. He was not an engineer. Nor had he worked on a space mission before. But he had studied Martian meteorites (using a mass spectrometer to analyse their chemical composition) and found organic molecules that were remarkably similar to those found on Earth.

'So, I went to the European Space Agency and said we have to take a lander and we have to do this experiment with a mass spectrometer,' Colin said with a broad West Country burr. It was a radical suggestion. Most mass spectrometers at the time were the size of a large chest freezer and just as heavy. Not the easiest instrument to adapt for space, where everything needs to be compact and lightweight. And they were notoriously tricky to operate in the laboratory, let alone at a distance of 93 million miles.

'I got the answer: "Who's going to pay for it? And who's going to do it? You can't build a spacecraft." One of the things you must not say to me is: "You can't do this." This is a fatal error.'

'Red rag to a bull, I guess,' Jim said.

'Yeah, my father always taught me there's no such word as "can't".'

When Colin was a child, his dad dug holes in the road for the gas board and fixed emergency gas leaks after hours, often taking Colin with him to help sniff out the leaks. His mother worked as a cleaner and used money put aside from her earnings to pay for Colin's school uniform so that he could go to Kingswood Grammar School, where he passed the 11+ exam. A week before his final undergraduate exams at Swansea University, he heard a talk about mass spectrometry by the Welsh chemist John Beynon (who pioneered using the technique to study organic molecules other than petroleum) and was enthused by the idea of being able to find things out for himself. How exciting to be able to work out what molecules were present in an unknown sample.

He went on to study the composition of Moon rock from the *Apollo* missions and several meteorites, using mass spectrometry. And in 1989 he wrote a landmark paper announcing the existence of organic molecules in a Martian meteorite. It was the first time such a claim had been made, creating some excitement and a lot of disbelief in the scientific community.

'It was very difficult to persuade people we had Martian rocks,' Colin said. 'But we did.'

Colin and his two new research assistants, Monica Grady and Ian Wright, were studying an 8kg lump of alien rock that had been found in Elephant Moraine, Antarctica, in 1979, and found evidence of rare gases and nitrogen isotopes that suggested a Martian origin.

The results of the next step were even more unexpected. They showed that the carbonates deposited in this Martian rock were very similar to the carbonates deposited by micro-organisms living in water columns on Earth. 'When we came along and said the carbonates have got organic matter in them, like carbonates on Earth,' Colin said, 'then that was really difficult.'

'It was evidence of possible life,' Jim suggested, perking up.

Colin sidestepped the question. 'It was really difficult

to persuade people of that,' he continued. 'And, of course, in 1996 some Americans – one of whom had worked in my lab,' Colin said with a wry smile, 'discovered what they thought was a fossil in the carbonate. Then there was an absolute furore, because NASA had to make an announcement. They thought they'd discovered life on Mars.'

'This is '96?'

'This is '96. August '96,' Colin confirmed. 'NASA was asleep when it was leaked. The BBC reported it and NASA's press conference had to be brought forward. Anyway, I realised then that I wanted to look for the organic matter.'

He found the evidence presented by the US team compelling: 'Some of it still haunts.' This was the moment he decided he needed to go to Mars, not wait to find more bits of Mars here on Earth.

The following year, the European Space Agency (ESA) announced that they were planning a mission to Mars in 2003 and Colin spotted his opportunity. The job of building a lander was put out to tender, and he put in a bid.

'We knocked out the competition,' Colin said, enjoying the victory all over again. 'After that we had to compete all the time with the fact that what we wanted to do was something extra to what the original mission was going to be.'

The European Space Agency had no intention of sending a mass spectrometer to Mars. Colin could not countenance going to Mars without one. And when it was clear that ESA would not provide the necessary extra funds, he looked elsewhere for support. 'Judith was the PR genius,' Colin said, referring to his wife.

Together Colin and Judith got artists, businesses and politicians involved, as well as scientists and engineers from all over Britain. Lord Sainsbury was a big fan, and generous too. The advertising agency M&C Saatchi and Saatchi got involved. As did the Queen. Damien Hirst created a special dot painting to calibrate the cameras. And Blur composed a six-note tune, programmed to alert mission control just as soon as the solar panels on *Beagle* 2 unfurled and it was ready to start work. 'It was my ringtone until my phone got run over!' Colin said.

Blur bassist Alex James explained how the band came to be involved: 'I called my accountant and said I wanted to start a space programme. And he gave me Colin's number.' The accountant lived next door to a satellite engineer who knew all about *Beagle* 2, and so Alex got to met Colin. 'I was instantly spellbound,' he said. 'He was just so instantly compelling and engaging ... He really had the X factor, all right!'

'We had Britart, Britpop, and now we have Britspace,' Alex told reporters back in 2003.

'It seems to me,' Jim suggested, mindful of the mountains Colin had moved to make *Beagle* 2 happen in the way he wanted, 'that the story of *Beagle* 2 really is the story of you refusing to take no for an answer.'

The Director of Science and Robotic Exploration at the European Space Agency, David Southwood, agreed. 'There's that spirit inside of him, that never-say-die attitude,' he said. Then he added with a hint of disdain, 'But you can't run everything that way. It's a great British tradition of strings and sealing wax, and once in a while we sort of get away with it. But if you want to win every time, you had better take precautions.

'We did our best to bend all the rules we could for Colin.

He won't admit that, but honestly, there wasn't anyone I knew working inside the European Space Agency who didn't want Colin to succeed. Or at least they never told me if they didn't.'

According to David, people at ESA were 'endlessly irritated by Colin's capacity to find a way round any argument'. 'But I told them: "He's a genius. We're working with a genius."' For six years Colin beat a path to David's desk, lobbying hard on behalf of *Beagle 2*. 'There are very few people like him,' David said. 'And sometimes I'm tempted to say, thank God!'

Colin responded immediately: 'All I can say is that they've had ten years with their bureaucratic procedures to get the next mission to Mars, and they haven't got one yet.' He sounded resentful. 'And it still isn't in sight by 2018. So who's right and who's wrong?'

Eight years after the event, it was as if Colin and David were having the argument all over again. 'I made it happen,' Colin said, defiant. 'Bureaucracy hasn't achieved it.'

Jim gently mentioned that David might have a point: 'When you're overseeing a very large project there are certain protocols and things you have to go through. You can't ride roughshod over the rules.'

'We went through every protocol, we went through every procedural review that they threw at us. They threw extra procedural reviews at us – we passed them. We didn't avoid any bureaucracy whatsoever. We provided all the paperwork. And I disagree with him: we didn't get total co-operation. Landing on Mars is difficult if somebody's actually not helping you.'

'Do you really think that they were somehow putting hurdles in your way for the sake of it?'

'I said that mission control should have been monitoring our signals as we descended. They decided they wouldn't monitor our signals. That, to me, was a criminal mistake. A major error. It could have been done. It should have been done.' Colin was becoming increasingly irate at this point. 'And if it had, we would have known an awful lot more about it.'

'Wouldn't they argue, though, that *Beagle* 2 wasn't the only project on *Mars Express* and they had others they had to be concerned about?' Jim suggested.

'I totally accept that they had lots of other people that they had to cater for as well. I totally accept it. It was an orbiter mission and we were the hitch-hiker,' Colin said. 'I accept that.'

Beagle 2 weighed just 69kg at the launch, which accounted for more than half of the total weight. 'We fought for every gram,' Colin said. He wanted to add extra systems to *Beagle* 2 and to make the systems they had more resilient, but with just 35kg available, it wasn't possible. ESA was not willing to allocate any more of the precious payload to *Beagle* 2, and it still irked Colin that David had other priorities. '*Mars Express* flew with 60 kilograms of additional fuel that we could have been using to make sure *Beagle* succeeded.'

'There were things that could have been done that weren't done,' Colin said defiantly. 'So, if I'm contradicting David Southwood, I'm contradicting him.'

Jim asked Colin if he did his best work when he was angry.

'That is absolutely true!' Colin exclaimed. 'I've thought of my best ideas when I've been irritated by people who mostly have done something which I consider to be silly or bureaucratic.'

'So, do you think it helped, having these bureaucrats at the European Space Agency to motivate you?'

'Look, it helped amazingly! It's the Sir Alex Ferguson syndrome. There's nothing like thinking the whole world is against you to get your team to play better.'

'Do you agree they bent the rules for you?'

'No, I bent the rules, they just had to—'

'I've thought of my best ideas when I've been irritated by people who mostly have done something which I consider to be silly or bureaucratic'

'Well, David Southwood seems to imply that they did everything they could to accommodate you.'

'I bent the rules and they had to accommodate them.'

'They turned a blind eye?'

'They turned a blind eye.'

Colin and his team clearly felt let down by ESA. 'My team always have believed that if we'd have landed on Mars, we wouldn't have got on the platform for the number of people who'd have been in front of us taking the plaudits.'

'Comments like the one from David Southwood about the *Beagle* 2 project being all string and sealing wax, that must hurt?' Jim said.

'That's not true!' Colin said. He was cross. 'I had the cleverest guys working in the British space industry and the universities building this. This was not sealing wax and string. This was top-quality engineering. I reject that claim entirely! It was not cobbled together. As I say, we tested everything.'

The scientists and engineers who were involved with

Beagle 2 routinely worked weekends, often staying up all night to meet a deadline or to try and solve a problem. Compared to their tireless efforts, ESA seemed to lack commitment. 'The problem was the budget,' Colin said. 'It's not that we didn't have enough. It's that we didn't have it quick enough. You can't come along a year before launch and suddenly throw 10,000 people at the job, because there're not that many people qualified to do those jobs.'

In the run-up to the *Mars Express* launch, there was some concern at ESA that *Beagle* 2 would not be delivered on time, a concern Colin quashed by volunteering to drive *Beagle* 2 to the launchpad in Kazakhstan himself, if necessary. Armed with this knowledge, Jim asked: 'So, it was a bit of a rush towards the end?'

Colin ducked the question. 'You have to be paid when you need the money,' he said. 'We were continuously asking for the money and were being told to jump through another hoop and then another. That isn't the way to do it.'

While ESA was managing Colin and his plans for *Beagle* 2, NASA was getting ready to send two more Mars landers of their own, *Spirit* and *Opportunity*. Colin found 'their softly, softly catchee monkey' way of doing things infuriating. 'NASA had this mantra that it kept repeating: "We're looking for the water and when we find water we'll know where the right place is to look for life." We didn't have the opportunity to think that way.'

'They were doing it by the book. You had one shot?' said Jim.

'I still believe that if you're a scientist and you're capable of doing the experiment, you should do the experiment,'

Colin said. 'And not try to make it into a long programme, because then you are going to be accused of just spending money to keep everybody in work. It would have frustrated me beyond belief to fly on *Apollo 10*: to fly so close to the surface of the Moon and then be told, "OK guys, home now."'

For six years Colin, Judith and the rest of the *Beagle 2* team worked relentlessly to build and promote a robot that was designed to look for life on Mars.

'We all knew that we were on a high-risk strategy,' Colin said. 'But the goal was big enough to merit the risk.'

Beagle 2 successfully separated from the speeding mother ship, *Mars Express*, on 19 December 2003 and entered the thin Martian atmosphere travelling at more than 20,000km per hour. Two days later the *Today* programme on Radio 4 reported: 'It's been another very frustrating day for the scientists involved in the *Beagle 2* mission to Mars. Their fifth attempt to make contact with the spacecraft has once again ended in failure.'

Colin was unperturbed. He was working his way down the fault tree to find out how the problem had arisen, helped by his highly motivated team of British scientists and engineers. Interviewed repeatedly about the likely fate of *Beagle 2*, he remained fired up with enthusiasm, despite the threat of possible humiliation. He responded to people's disappointment by spreading hope and described the suspense as like sending someone a love letter and waiting for an answer.

'At what point did you think, oh dear, this is no good?' Jim asked.

'I never, ever believed we had crashed,' Colin said. 'And I certainly don't now. Something didn't work,' he said, a broad smile emerging. 'The most annoying thing is that we don't have the sequence of events recorded through the radio beacon that we were intending to put on *Beagle* 2.'

A month later, ESA announced that *Beagle* 2 was officially lost in space. Colin continued to search for it, nonetheless. Jim wondered if it felt a bit like the family cat going missing. 'You sort of hold out hope for days that it's going to wander back in, bedraggled. And at some point, you just have to accept that it's not coming back.'

'I'm used to the family cat going missing,' Colin said, laughing. 'We've had lots of cats that have wandered off, but that was the thing about the name,' he said, steering the conversation towards a familiar anecdote. 'At the outset everybody had thought the name *Beagle* 2 was wonderful.' (It was chosen by his wife Judith in memory of the vessel that launched Charles Darwin as a naturalist.) But when *Beagle* 2 went missing, their feelings changed. 'We had all manner of emails, texts, telephone calls, letters saying, "Don't you realise that Beagles are the worst dogs you could possibly let off the leash? They run off. They chase things. They don't come back when they're called. They only come back when they're hungry and they show no sign of remorse."'

'And so it probably fits quite well . . .'

'It fits absolutely perfectly!' Colin laughed. '*Beagle* 2 was the inspirational name that people could equate to it. They knew what we were talking about straight away.'

'If you were to dissect the Pillinger pie,' said Roger Highfield, a science journalist who covered the *Beagle* 2 story extensively, 'you'd find you've got 50 per cent really good scientist there. You've got 25 per cent of someone with an amazing drive and relentlessness. And 25 per cent real showman coming up with clever ways to promote what he thinks we should all be doing. And it's very unusual to have someone in the scientific community who's got all those traits in one package.'

'Did the showman ever get in the way of the scientist?' Jim asked.

'No, I don't think I was a showman,' Colin said. 'My wife was very good at dreaming up ideas. I was the person who delivered them. I've often said this was an event which was waiting to happen. I just happened to be the pied piper who was there to lead the people. They were wanting to do this.'

'But you do have a certain common touch,' Jim said, 'a skill in inspiring and exciting the public about what you do.'

'I have a common touch because I came from a common background!' Colin said with a mix of pride and outrage. 'I know the audience. I don't do public understanding of science for people who are converted already. I'm not talking to people who watch *Horizon*. I'm talking to the taxi driver and the man I stand next to in the pub or the checkout at Sainsbury's. It's no odds to me who I try and tell about science. I find science fascinating. It's marvellous. It's full of surprises. And I love surprises. You find out things that you and nobody else has ever known. For a few minutes, hours, days, you're the only person on the planet who knows this information, and then I just want to share it with somebody.

'Every generation needs its Gothic cathedrals. We shouldn't be doing things that everybody else has done

'It's no odds to me who I try and tell about science. I find science fascinating. It's marvellous. It's full of surprises'

already. We have to do something which is adventurous and risky. We have to do something we don't know the answer to, because if we know the answer, then somebody else has done it already. It inspired the whole nation. There were people sitting on the edge of their seats on Christmas morning . . .'

'I was one of them,' Jim said.

There were kids, Colin was told, who got up on Christmas morning 2003 and asked not, 'Where are my presents?' but 'Has *Beagle* 2 landed?'

'In a sense, then, do you see that the *Beagle* project wasn't really a failure?' Jim asked.

'The word "failure" does not exist in the language of anybody that worked on *Beagle*.'

'I mean, it didn't do what it was supposed to do in that sense, scientifically,' Jim suggested politely.

'We learnt an enormous amount out of *Beagle*. It was not a failure,' Colin replied firmly.

'Is there anything you'd have done differently?'

'Yeah, I would have made landing on Mars the priority,' Colin replied. 'If I ran this experiment again, landing on Mars would be the priority.'

Colin died in May 2014 from a brain haemorrhage, having suffered from MS for many years. Six months later, the HiRISE camera on NASA's Mars Reconnaissance Orbiter *detected something unusual on the surface of Mars. Closer inspection revealed that it was* Beagle 2, *lying on its side, not far from its intended destination.*

Colin was vindicated. Beagle 2 *was not lost in space – it was only the communication system that had failed, just as* Colin *had always said. Soon after, Colin's friends and colleagues gathered to remember him and celebrate* Beagle 2*'s arrival, eleven years after the event.*

Alex Mackworth-Praed

HENRY MARSH

'The inside of a living brain is a slightly firm white jelly'

Grew up in: Oxford

Home life: married to the anthropologist Kate Fox, with three children (William, Sarah and Katherine) from his first marriage

Occupation: neurosurgeon

Job title: Senior Consultant Neurosurgeon at St George's Hospital, London

Inspiration: witnessing a technically exquisite emergency brain operation

Passion: neurosurgery

Mission: to encourage honest conversations about death and to encourage doctors to talk about their mistakes

Advice to young surgeons: talk to each other

Date of broadcast: 30 June 2015

Henry Marsh was the first person in the UK to operate on someone's brain and chat to them at the same time, when he pioneered brain surgery under local anaesthetic.

For many years, he was unsure what he wanted to do with his life. He ran away from Oxford University for a year to be near the girl he loved and he decided to study medicine relatively late in life, having done very little science at school. Witnessing 'a technically exquisite' aneurysm operation convinced him that he wanted to be a neurosurgeon, his son William having been diagnosed with a brain tumour a year before. Good decision making, he insists, is as important as an ability with the scalpel, and he is determined to encourage surgeons to learn from their mistakes.

He talked to Jim a few months before he was about to retire from St George's Hospital in London, where he worked as a consultant neurosurgeon for more than 30 years.

The essence of our humanity resides in just over a kilo of 'slightly firm white jelly'. A shiny globule, not much bigger than a grapefruit, generates our every thought and feeling.[1] It contains our genius and our passion. So what, Jim wondered, does Henry Marsh think about when he cuts through our brain jelly with a scalpel? 'It must be unimaginably stressful.'

'I'm always nervous before operating, but it's a nervous excitement and not just nervous fear,' Henry said, with disarming candour.

'Is there a moment when you switch into surgeon mode?' Jim asked. 'When you first approach the operation . . . ?'

'I do a bit,' he said. 'I usually visit my patients the night before an operation when I'm still Dr Jekyll, so to speak. I'm not exactly Mr Hyde next day, but I'm certainly aware of the fact that, psychologically, I don't really want to have to bother with being nice to the patients immediately before the operation . . . I do it but I find it more difficult.' I just want to wear blinkers.' It's not so easy to chat to someone when you're about to cut open their head.

'The great problem in medicine is finding a balance between detachment and compassion,' Henry said. 'And it's a process . . . you have to learn to cut yourself off a bit . . . In fact, you can get too involved in patients. I mean, I did

---

1 A typical human brain weighs between 1,200 and 1,400g, about the same as a walrus brain. Dead brains, pickled and stored in jars, look 'like dirty old giant walnuts', Henry said. 'They're disgusting.' But the living brain, under the microscope, looks wonderful. The spinal fluid shines brightly through an exquisite filigree of blood vessels.

once operate on a very close friend with a brain tumour. A very simple biopsy operation. And even that was almost impossible because I was so bloody nervous.' Dr Jekyll would not leave Mr Hyde alone.

Many doctors decide when they are quite young that they want to save lives, but Henry took his time. For many years, he was very unsure what he was doing with his life. He studied Philosophy, Politics and Economics (PPE) at Oxford University but 'ran away' after two years, 'for various silly reasons such as unrequited love – well, sort of unrequited – and a sort of nervous breakdown.'

'I've been very successful in my career,' he said, continuing seamlessly. 'And we all like to attribute our success to our efforts. But honestly, in my career – yes, sure, I'm jolly clever and all that – but actually, a huge amount depends on luck and the kindness of others. And also being in an advantageous position. Being from a middle-class, educated family background is very helpful.'

'Is it some sort of sense of entitlement that, yes, I can do this if I want to?' Jim asked.

'My second wife, Kate Fox the anthropologist, is always saying, "You public school types!"'

'She's envious of all this self-confidence you have . . .'

'Which you do get, but it's a pity everyone can't have it.'

Having left university rather abruptly to follow the girl he loved, who had only 'sort of' loved him back, he spent six months working as a hospital porter in Ashington, a large mining village north of Newcastle, wheeling people in and out of operating theatres. He had no particular desire to become a doctor, at this stage. The job was simply a means to an end. He then spent another six months organising conferences at the School of African and Oriental Studies in London.

'During your time away, did you have any sort of plan for your future?' Jim asked.

'No. None at all. I had sort of reached rock bottom and I was very uncertain what I'd do.'

After this unscheduled year off, he returned to Oxford. 'Although I had behaved very stupidly, my Oxford college kindly allowed me to come back,' he said. He worked hard and graduated with a First in PPE, then decided he wanted to be a surgeon. 'I always liked using my hands. I'm always making things. I have a very large workshop where I make furniture and all sorts of stuff. And I thought surgery would combine hand-work and thinking.'

Fortified by a middle-class sense that if he wanted to do something enough, then there must be a way of achieving it, he applied to medical school and enrolled on a one-year crash course in A Level physics, chemistry and biology. Science 'hadn't been very well taught' at his expensive school but he found this course 'absolutely fascinating and was not bored by it at all'.

He later trained at the Royal Free Hospital in London and became a junior doctor. Then doubt set in. 'There were two very nice general surgeons who were terribly kind people, really nice. But, on the whole, when I was doing general surgery there were long incisions and smelly body parts . . . And I didn't like it very much.' Inside the operating theatre, it was gruesome. As a hospital porter he'd only seen the before and after, not the during. 'There was something almost offensive about it,' he said. 'So, I got rather depressed.'

Yet more indecision loomed. Still he 'went on training' and, a few years later, was rescued. Working as a senior house officer, helping to administer an anaesthetic in intensive care, he witnessed an operation that was life-changing both for the patient and for him. 'Memory can be deceptive and I hope I'm not exaggerating,' he said. 'But I like to tell the story as an epic experience. I saw an aneurysm operation on somebody's brain. And I came home and told my first wife, I'm going to be a brain surgeon. Full stop. And I've never regretted it. It's often very painful and very demanding but I think it's absolutely fascinating.'

'And I came home and told my first wife, I'm going to be a brain surgeon. Full stop'

'What was it about that operation in particular?' Jim asked.

'The glamour of it all,' Henry replied without hesitation. 'And the danger.'[2]

The operation was 'technically exquisite'. An aneurysm is a blow-out in a blood vessel in the brain. 'It's less than a centimetre in size and ready to blow up at any moment.'

'Imagine the tube of a bicycle tyre,' Henry explained, as he does to his patients. If there's a weak patch when you pump up the tyre, 'it will start to bulge out like a little swelling.' Then it will turn into a balloon and 'one day it bursts'.

Typically, aneurysms occur deep in the brain in the spaghetti junction of blood vessels known as the circle of Willis. The surgeon's job is to seal off the pushed-out bubble in the cerebral blood vessel. It's tiny and pulsating and the consequences of not acting quickly enough can be lethal – a pressurised jet of blood spurting forcefully

2 Aneurysms are always treated as medical emergencies. Mercifully, they are quite rare.

through a hole in the wall of the blood vessel, buried deep within the brain.

'It was very dangerous surgery, it was very delicate surgery, very exquisite.[3] And all done on an operating microscope. It appealed to my sense of self-importance. But it was also to do with the brain. And it was terribly interesting.'

Henry's son William was diagnosed with a brain tumour when he was just three months old and was operated on, successfully, a year before Henry's epic conversion. 'Did that influence your decision to become a neurosurgeon?' Jim asked.

'Well, it must have,' Henry said, 'but not knowingly . . . We tend to be very binary. We always think, if *p* then *q*, single causes and realities, not lots of things together.' He paused for a moment. Then concluded, 'I'm sure it must have.'

'And, of course, they say doctors make the worst patients. In this case it was your son who was the patient . . .'

'Yes. It is very difficult being a medical relative,' he said, spinning away from the personal to make a general point. But 'it's very important. One of the big problems in healthcare is that most of it is being doled out by fit, young, healthy doctors and nurses who just don't understand what it's like to be a patient. And how miserable it is, often. Like being in prison. I don't know what sort of doctor I would have been if I hadn't had that experience. But I'm sure it made me a better one than I would have been.'

The only time he was a patient himself, he had a retinal

---

3 Aneurysms are rarely treated like this today. Instead, by means of keyhole surgery, surgeons enter into the brain inside blood vessels and mend the puncture from the inside of the tube. There is no longer any need for invasive surgery.

detachment, 'first in one eye, then in the other. And it is fairly serious surgery. It was like having four very large needles stuck in your eyeballs, and they then suck out the tumour. But waiting in the anaesthetic room for the operation, I realised retinal surgery was nothing compared to what I was doing to my patients. I had no right to fret, at all. And I didn't. I was very surprised. I've always thought of myself as a coward. I am a coward and I'd always dreaded being a patient myself.' Much to his surprise, however, he realised that he had learnt something from 'seeing so much evil misery at work'. 'I was lucky to have a problem that was fixable,' he said. 'And my eyesight was saved.'

'I'd always dreaded being a patient myself'

When the Scottish explorer Mungo Park marched through West Africa in search of the source of the river Niger in 1788, geography was a new science. He ventured into the interior with a compass and a rudimentary map and stuck as closely as possible to the river, for fear of getting lost. He knew little about the territory he was passing through and proceeded boldly into the unknown, single-mindedly pursing his goal.

A neurosurgeon entering a vast expanse of brain jelly might feel the same. There is still so much we don't know about the human brain, despite the rapid advances in neuroscience in recent years. We have created regional maps. Eloquent parts of the brain are so named because they facilitate speech and help to control our arms and legs. But for a brain surgeon going in, most brain jelly (eloquent or otherwise) looks the same.

The more we learn about the human brain, the more

subtle and complex it becomes – a network of a hundred billion neurons, an intricate and ever-changing mesh of interconnections – and consequently neurosurgery seems 'increasingly crude'. Good operating microscopes and computer-assisted navigation help, of course, and fine tools facilitate some delicate stitching up. But 'it is ultimately destructive'. All a neurosurgeon can really do is chop a bit out.

It's easy enough to spot a brain tumour, and it's not so hard to cut around it and remove it: 'easier than woodwork', Henry said. 'In woodwork you have to get the joints right first time. Brain jelly is more forgiving.' But the challenge is to avoid damaging surrounding brain cells which could perform a vital function. Tumours are best removed, but not at any cost. A tumour might grow slowly. Damaged nerves, on the other hand, can destroy a patient's quality of life in an instant. Depending on where the tumour is located, chopping out too many surrounding cells can result in disability, paralysis or death. A tumour growing close to a major artery is best left untouched.

A neurosurgeon moving a scalpel towards a tumour may know little about the brain he or she is passing through. All nerve pathways look the same. Rather like having a road map which fails to distinguish between B roads and motorways, it can be hard to know if the nerves that are being destroyed are fundamentally important or little used.

Henry, however, came up with an ingenious solution to the problem of not knowing which brain cells were essential and which could be sacrificed with little consequence. Keep the patient awake. He was the first neurosurgeon in the UK to remove a tumour from a patient under local, not general, anaesthetic. By getting the patient to chat, wriggle a finger or raise an arm, he could monitor brain function while he cautiously checked the terrain. An alert patient can help to guide a surgeon's scalpel and limit damage to vital areas of the brain. At the slightest sign of difficulty, Henry would know to stop immediately. (If only a few cells are damaged, there is a good chance they will grow back.) For as long as the patient was interacting normally, he had a licence to proceed.

It was not risk-free, but it made it possible to remove tumours that were located in areas that were previously considered too much of a gamble. By proceeding with caution and checking brain function, the risk of surgery-induced disability could be reduced.

'Just the thought of someone digging around in my brain is enough to make me panic,' Jim said, 'let alone being awake during the operation itself. So, you know, how does that feel for the patient?'

'Well, I've done this over 400 times, and only twice could the patients not cope with it,' Henry replied, matter-of-factly.

'But they must hear the drill going into their skull?' Jim said, concerned.

'That's true. We have them under general anaesthetic for that part of it, awake for the mapping and removing the tumour, and then asleep for the stitching up and the end.'

'Do you ever worry that they know what's going on . . . ?'

'I'm talking to them and they'll say, "Go on Henry, take it all out!" That sort of thing.'

'And what about if something maybe goes slightly wrong? Do they pick up on your stress?'

'Actually, this particular sort of tumour surgery is fairly safe. If you think of the brain like a tree, you're operating out in the twigs, you're not operating near the major arteries or the tree trunk.'

Of all the neurosurgeons Henry has trained, very few have failed because they couldn't master the mechanics. It's the decision making that's hard. Weighing up the risks associated with not operating (from a brain tumour which may be growing unpredictably, for example) against the risk of disability or death caused by the surgery itself, is not an enviable choice.

'Was there a moment for you when you realised that sometimes the best course of action is to not operate?' Jim asked.

'In neurosurgery, knowing when not to operate is hugely important. I mean, it's an old cliché, but it takes three months to learn how to do an operation, three years when to do it, and thirty years when not to do it.'

'Because what we're talking about here is a patient living with a brain tumour, and then weighing that against the risk of surgery, which may lead to paralysis . . .' Jim suggested.

'The question is: is it worth prolonging that person's life? Are you prolonging dying or are you prolonging living? And these are very difficult questions to answer. And they're very difficult for patients and their families to answer . . . It's not cut and dried at all.'

'And on that point of deciding not to operate,' Jim said, 'that must be very difficult for a surgeon, who's pretty much programmed to try and save lives.'

'Absolutely. It's very difficult. It also means you cannot escape a long and difficult conversation with a patient and the family. And that's the difficult bit. To say, "Go away and die" is incredibly difficult. And there are all sorts of mechanisms we use to try and keep these problems at arm's length. It's always easier to treat, rather than not to treat.

'When I was on call, I'd be rung at night by my juniors with emergency cases – head injuries, cerebral haemorrhages, things like that. And I'd have to make a sort of rough judgement – is it worth saving that person's life, or not, by operating? And if I say, "Operate," I go back to sleep. If I say, "No, let them die," I find it very difficult to go back to sleep, because I'm worried I might be wrong . . .

'The great problem with modern medicine, neurosurgery, is: we actually generate a lot of human suffering by keeping people alive . . . And it's very difficult as a doctor. Do you tell the family, "In my opinion, I think your nearest and dearest is better off dead"? I could get away with that because I am old and senior. And have been there and I've got grey hair. For a junior doctor, working nights, in an emergency, it's very difficult.'

'We actually generate a lot of human suffering by keeping people alive'

Henry has saved hundreds of lives by operating on people's brains, but the thing that he is perhaps most proud of is the regular Monday morning meeting he set up with his group at St George's Hospital in London, to get neurosurgeons talking to, and learning from, one another. Many medical students find it terrifying, perhaps because they are forced to open themselves up. It can feel like failure, to a high-achieving medical student, to acknowledge doubt. You don't get marks in medical school for not being sure.

'We've got to get away from the monotheistic religion of the single surgeon doing everything,' Henry said. 'The answer is to work together and share experiences. It's not collective decision making. It's rather that we should discuss cases together before we do things. That, I think, is the solution. You need a bit more honesty and openness.' Especially when things go wrong. 'I've got very nice colleagues. We all get along very well together, and you know when somebody has had a disaster or a bad result. One rolls one's eyes. We know what it's like. We've been there. But it's not something you talk about at great length.'

'A lot of people accuse surgeons, neurosurgeons in particular, of being arrogant or cavalier, or both. Do you think that's fair?' Jim asked.

'No, I don't . . . We're humble because we know we do make terrible mistakes sometimes. And when we make mistakes, it's awful . . . You'd have to be a real psychopath for it not to hurt. But you know what you're doing is terribly serious. And very important. And it can sometimes manifest as being arrogant and self-important. Most of the neurosurgeons I know are not. They're mainly slightly tortured. And they have to hide it.'

'The stakes in neurosurgery are so high, the outcomes are far from certain, and yet you always have to appear calm and confident in front of the patients and their families,' Jim said. 'That can't be easy . . .'

'No, it's not. You just have to do it. You have to act. There's nothing more frightening for a patient than to see an anxious, frightened doctor who's not certain what to do.'

'There's nothing more frightening for a patient than to see an anxious, frightened doctor who's not certain what to do'

'How do you help patients to decide on the best course of action when they are so dependent on your knowledge and expertise?' Jim asked.

'I always encourage patients to get a second opinion if they have doubts. The irony is, of course, that by saying that, they immediately trust you.'

'Perhaps it's inevitable that patients are going to put you on a pedestal and you become either a superhero or a villain,' Jim said.

'Yes. And it's something to resist.'

'How, then, do you bring yourself back to earth?'

'Just remembering all the mistakes I've made,' Henry said. 'And worrying all the time about it. I mean, it's a balance. And these are not things that surgeons talk about among themselves.

'The French surgeon René Leriche says all surgeons carry within themselves a cemetery, and it's a place where they must go to from time to time to contemplate. And it is a place of bitterness and regret ... Being a senior surgeon, my cemetery is quite large,' Henry said, plainly. 'And it's something that is just part of my life. It's just there. But it also guides me in my decisions. And although I'm retiring, so it's no longer very relevant, it is relevant to the extent that I'm going to go on teaching. I'm going to go on working abroad. And the critical thing is to help my trainees not to make the mistakes I made.

'I've tried very hard to remember as many of my mistakes as possible. It's extremely difficult.' Before he wrote his first book, *Do No Harm*, he would lie in bed every morning, forcing himself to access painful memories, and he found that if he didn't write them down immediately, he would forget them all over again. 'I'm pretty sure that probably some of the worst mistakes – I'm talking careless mistakes – are still buried. But the more I think about it, the more things keep coming back.'

'Are there particular operations that continue to haunt you to this day?' Jim asked.

'I'm not haunted, because I do take the view that the

triumphs are the triumphs because of all the failures and mistakes.'

'I do take the view that the triumphs are the triumphs because of all the failures and mistakes'

Mistakes need to be discussed openly so surgeons can learn from the worst practice as well as the best. 'And I feel very strongly: I am a human being and I make mistakes. And I think the practice of medicine would be better if that was generally more accepted, both by doctors and by patients.

'We've got to get away from the idea of the surgeon as a solitary, Michelangelo-style genius . . . I think those days are over now. We've got to change the model.'

Brain surgeons are mere mortals, after all. Armed with little more than an operating microscope and a scalpel, they venture into the unknown.

'Will you miss the operating theatre?' Jim asked at the end of the interview, aware of Henry's imminent retirement.

'Yes and no. I won't miss the stress and anxiety,' Henry said. 'And the other problem is, I do find brain surgery increasingly crude.' It's hard to read about the latest advances in neuroscience and then hack away at the subtle complexity of the brain.

'Will you miss the euphoria?'

'No, because I stopped feeling euphoric a long time ago,' Henry said. 'It's much more a sense of relief rather than glory, because there have been too many disappointments in the past . . . It's not false modesty: I'm just hugely relieved if the patient has survived and done well.'

Courtesy of Hazel Rymer

HAZEL RYMER

'I measure gravity'

Grew up in: Reading
Home life: married to chartered engineer Carl with two boys, Ben and Sam
Occupation: volcanologist
Job title: Professor of Environmental Volcanology, The Open University
Inspiration: visiting an exhibition about Pompeii
Passion: measuring microgravity on volcanoes
Mission: to anticipate volcanic eruptions and so help save lives
Best moment: 'When the penny dropped and I finally understood the plumbing in the Poás volcano in Costa Rica'
Worst moment: 'Being on the rim of an active volcano when an earthquake struck'
Advice to young scientists: 'Follow your passion'
Date of broadcast: 28 June 2016

Hazel Rymer is a volcanologist, a physicist who turned her attention to the physics of the Earth. Early in her career she pioneered a way of 'seeing' beneath the surface of a volcano, using tiny changes in the Earth's gravitational field in and around the crater to detect when magma is on the move. She has studied several volcanoes in great detail, looking for patterns and seeking to understand the events that precede an eruption. Initially she was fascinated by the mechanics of volcanic explosions. More recently she has spent a lot of time trying to mitigate their impact. Thousands of lives could be saved every year, if only we could predict when individual volcanoes were going to explode and evacuate residents in time.

Most of us (most of the time) forget that the Earth is on fire beneath our feet. We live on the surface, blissfully unaware of the superheated ball of pent-up energy at the centre of our planet. Volcanoes seem to explode in an instant, and at random. Sometimes boiling mud-pools appear on the surface. The ground may shift, a little. But typically, there are no obvious signs of an imminent eruption.

'Every time you visit a volcano it's different and it's just stunning – the power that is there,' Hazel said.

'It's that sense of raw, elemental power under your feet . . .' said Jim.

'Yes. What draws me to volcanoes is the fact that we can't do anything about them . . . We spend so much of our lives trying to fix the environment, or fix stuff around us and make it how we want it – volcanoes are not having any of that! They just do what they want. Even a small volcano you can't put a stopper in it.'

'I guess they do bring home the fact that we are still very puny when faced with the forces of nature,' Jim said.

'Absolutely, and it's wonderful' Hazel replied.

When Jim suggested that spending time on, and sometimes in, volcanoes was optional, Hazel replied as if it wasn't. 'The risks add up. Of course they do,' she said. 'Bad things do happen, but bad things happen at home and in the workplace too. I always think that it's far more dangerous driving around on the motorway to the airport and flying out of it than it is doing fieldwork on a volcano.'

'And when you go to a new volcano and want to get the best possible data,' Jim said, 'do you find yourself desperately wanting to get as close as possible to the inside of the crater, and at the same time knowing that it's probably not a very clever thing to do?'

'It all depends on why you are at the particular volcano. I'm making these measurements in order to understand more about volcanoes. So getting a measurement right down inside the crater, if that's going to help with my research goals, then fine, I'll try and do it. But not if I'm not going to make it back. Or if I'm going to break the instruments, because that defeats the object.'

'Do you have moments when you think, "Shall I, shan't I? I'd like to put the meter there, but . . . ?"'

'Yes, you do have moments like that. The whole time you're assessing this, of course. But you know this is not about trying to lose people or equipment. It's about trying to collect data.'

'Nevertheless,' said Jim, 'people have been killed carrying out measurements inside volcanoes, haven't they?'

'Yes, many times,' said Hazel. 'The time that affected me most significantly was back in 1993 when one of my PhD supervisors was killed in a volcanic eruption. They were inside an active crater, making gravity measurements, and there was a completely unexpected eruption. It was a very small eruption, but because they were close by, they were completely lost and have never been found.'

'How easy was it for you, after that event, to get back to what you were doing?'

'Well, I still say, if someone gets killed in a car crash, do you not get in a car the next day? . . . The explosion had not been expected, that was the really, really scary thing. That sort of eruption is, to this day, very hard to predict. There are seismic signals, little earthquakes that you can see. With hindsight you can look back through the record and say, "Well, OK, there was some sort of precursor." We know about these precursors now, but often they don't lead to an eruption. It was just really very, very unlucky that they happened to be so close to where the explosion took place. It was a horrible tragedy.

'Does it make me think it's not worth trying to understand how volcanoes work, and why they work, so that we can predict, not those smaller explosions, but much larger ones that will impact on communities living near volcanoes, and even global climate? Absolutely not. I think the science needs to be done.'

Hazel discovered geology when studying physics at Reading University, happily swapping long days in the laboratory for fieldwork on volcanoes. 'You can do anything once you've got a bit of physics behind you,' she said.

> *'You can do anything once you've got a bit of physics behind you'*

Geophysics was a good way of combining 'two very different ways of thinking'. As a physicist Hazel had measured everything precisely, calculating error bars and making all unknowns, known. Geophysicists, she discovered, spend more time estimating and extrapolating, humbly trying to gain insight: servants, not masters, of the universe. 'Sometimes you don't even know what the uncertainty is,' Hazel said, and for a physicist who had always aspired to absolute accuracy, this came as quite a shock. She soldiered on, taking measurements as precisely as she could, in the most uncertain and unreliable conditions. And, early in her career, she found a way of monitoring what was happening underground, by measuring tiny changes in the Earth's gravitational field.

Gravity, as Newton discovered, is a universal force, responsible for holding our solar system together and making apples fall on heads. It keeps our feet on the ground, but it is easily overcome. (Just jump!) When an apple (or any other object) falls, it hits the ground at quite a speed, depending on the distance travelled. Gravity causes it to accelerate at 9.8 metres per second squared. But that number is an approximation. Measured a million times more accurately (to six decimal places, not one), it varies.

Hazel uses hyper-accurate measurements of gravity to draw contour maps in which the lines represent different microgravity readings. From these maps she is able to build up a picture of the rocks below. This works because the microgravity at a particular location tells us about the density of the rocks beneath the surface. High-density rocks have more gravitational pulling power than, for example, pumice, which is full of air. Magma, typically, is considerably less dense than the surrounding rock. (It's not always the case but that's the general rule.) And so, rather like a plumber tapping walls and listening for different sounds, Hazel searches for volcanic pipes, using micro-variations in gravity as her guide. In particular she is interested in any changes in microgravity that take place in the same location over time. Rock density (the mass-to-volume ratio) generally remains the same, and so when the density at a particular location looks as if it has changed, it suggests that either the composition of the rocks has changed or that there are now different rocks in the same location. The most likely explanation of either of those phenomena is that magma is on the move, finding and creating fissures as it rises to the surface, heating nearby rocks and changing their mineral composition or moving things along.

'If a pipe is feeding an active crater and the density within it changes, then you can start to model what might be going on.' And in this way, Hazel hopes to be able to detect if a volcano is getting ready to erupt.

The instruments Hazel uses to measure gravity look rather like car batteries and are just as heavy. (Not everyone enjoys lugging them up volcanoes.) They look robust but are finicky and awkward to use in the laboratory, let alone on the side of volcanoes. 'Electronics and the outdoors don't always work very well together,' Hazel said. 'Gravity meters are supposed to be used for fieldwork but they don't like rain and they don't like wind. And they don't like volcanic ash landing on them because you can't see the numbers on the top.'

Not ideal. But Hazel seems happy to live with these foibles and has remained loyal to her equipment. 'I've been with one of my gravity meters for nearly thirty years.'

'It sounds like there's some affection there.'

'Absolutely! Yes. I love gravity meter 513.'

'I've been with one of my gravity meters for nearly thirty years'

'It sounds like there's some affection there'

'Absolutely! Yes. I love gravity meter 513'

It's 'a very simple but rather expensive and delicate piece of kit', a mass on the end of a spring. When the Earth's gravitational pull increases even very slightly, the spring extends. The biggest challenge is making sure that any changes to the spring are caused by the acceleration due to the gravity of the Earth, and not other things – a moving arm, for example, or a colleague passing by. (Most of her colleagues now know to steer well clear when Hazel is taking measurements.) To accurately track changes over time, Hazel needs to be sure she is comparing like with like. She marks the spot where she's taken a measurement with pins and paint, and when she returns, months or years later, she places GM513 in exactly the same position, sitting facing in exactly the same direction, cross-legged and with her back to the prevailing wind. 'There's a particular way I sit, and I have to do it every time because everything has to be as similar as possible.' The smallest things can make a difference when you're measuring microgravity.

'Whereabouts do you put these gravity meters?' Jim asked.

'A good gravity station is one that I can find again the following year, for a start. In an ideal world you would get right up to the closest part of where you think the magma is. So if that's down in the crater bottom, then we try to get down there to take measurements.'

For a long time, scientists studied volcanoes because they could. 'It was just a cool thing to do,' Hazel said. When she first became interested, her overriding priority was to understand how they worked, but more recently she has been thinking about the people who live near them. Understandably, but unhelpfully in this respect, the most well-studied volcanoes are rarely the ones that pose the greatest threat to human life.

'Science is all very well but it needs to get out of the ivory towers'

'Science is all very well but it needs to get out of the ivory towers,' Hazel said. Tens of thousands of lives could have been saved in the last few years if we had known in advance when each of the volcanoes that have erupted were going to explode. 'It's fantastic to understand the nuances and tiny things within magma feeding systems, but if I can't explain why it's useful for me to go out there and make those measurements, perhaps it's not useful.'

'So then, I have to ask: how close do you think we are to being able to protect people from the most devastating impact of volcanic activity?'

'We can't stop a volcano erupting. We can start to mitigate some of the impact, though.' By understanding the changes that are taking place underground in the run-up to a volcanic eruption, Hazel hopes to help to develop early-warning systems for people living near volcanoes.

'And presumably you can evacuate people in advance and protect lives?'

'Indeed. And that's the really interesting thing. There are so many volcanoes that are not monitored at all at the moment. With instruments becoming cheaper and cheaper, we should be able to monitor more volcanoes more of the time. And, of course, satellite technology is helping a lot with that . . . Being able to see some precursors and being able to evacuate people in a timely manner is one thing that's fantastic.'

Volcanologists are getting better at spotting patterns, and the physical processes that precede volcanic eruptions are now better understood. As a result they are sometimes able to predict eruptions, but just as early-warning systems are starting to be developed, so scientists are realising that there's another pressing question that needs to be answered if nearby residents are going to be kept safe: 'People really want to know when the eruption is going to stop, and actually that's at least as hard. If not harder.'

How long does a volcano that's erupting need to vent

its fiery fury: a week, a month, a year? People can normally be persuaded to evacuate in an emergency. They are, however, always keen to get back home.

Decades after she developed them, Hazel's microgravity measurements remain the only way to spot ground-mass changes. But the study of volcanoes has moved on. These days, volcanoes are monitored remotely using radar images taken from satellites in space. Gravity-measuring devices have been transformed by the digital revolution.

Hazel's gravity meter 513 and others like it might soon be replaced by armies of cheaper, lighter, smarter instruments. Hundreds of tiny digital devices could be dropped from drones into locations too dangerous for humans, where they can continuously take measurements and be monitored from the comfort and safety of an office. With the right protective shield on these digital devices, a system of data collection could be put in place that would be risk-free.

Jim wondered where this would leave Hazel and gravity meter 513. She replied with dignity: 'All volcanologists have to put away their boots at some point.' But there would always be a need for ground truth. 'You can't replace physically being there with remote sensing. You've got to have a deep understanding of what's going on.' And despite all the exciting new technology that's coming on stream, she hopes and believes there will always be a need to do things the traditional way: with hiking boots and a heavy black box on the rim of a volcano.

HELEN SHARMAN

'I didn't go into space to be famous'

Grew up in: Sheffield
Occupation: astronaut
Job title: Helen is now an operations manager at Imperial College, London
Inspiration: hearing a radio advertisement: 'Astronaut wanted'
Passion: following a process from beginning to end
Mission: to do experiments in space
Best moment: looking back at planet Earth
Worst moment: being bombarded with personal questions by journalists
Advice to wannabe astronauts: 'Normal kind of people can do amazing things'
Date of broadcast: 15 March 2016

Helen Sharman is the first British person to go into space. Driving home from work one day when she was 27 years old and working in a food factory, she heard an unusual advertisement on the radio: 'Astronaut wanted. No experience necessary.' The last leader of the Soviet Union, Mikhail Gorbachev, wanted a British astronaut to fly with the cosmonauts to the Soviet space station *Mir* to demonstrate his commitment to *glasnost*. Helen was astonished to be selected. She was sent to a military training base just outside Moscow to undergo training for 18 months, and on 18 May 1991, she strapped herself into a *Soyuz* TM-12 capsule in south Kazakhstan and was blasted into space.

Jim interviewed Helen a few months after the British astronaut Tim Peake had left Earth. The media coverage of his departure had been intense. Images of his every move on the International Space Station were beamed into our sitting rooms night after night for weeks. In December 2015 the headlines shrieked: 'First Brit in Space'.

It was not true. Helen Sharman, not Tim Peake, was the first British person to go into space.

So how did Helen feel about the fact that everyone seemed to have 'rather lost sight' of what she had done? 'How did it feel to be erased from the nation's memory?' Jim wondered.

Helen hesitated briefly. 'It was great to hear so much more news about spacefaring. And in a very positive way.'

'I mean, the coverage was a bit misleading, at times . . .' Jim said.

'I think the press had been given a certain steer – let's say from the UK Space Agency a few years ago – that this might be the first British astronaut in space, and they were repeating things that they had heard before.'

'There must have been a little bit of you that was thinking, "Hang on a minute. Hello? I'm still here!"'

'I don't think people overall had forgotten that I'd flown. But I mean, even if they had, I'd been out of the spotlight for many years and public memory is actually relatively short.'

'If that was me, and people talked about someone else being the first Briton in space, I would be put out!' Jim said, outraged on her behalf.

Helen remained gracious. 'I was more than happy to do

a number of interviews at the time,' she said stoically. 'So I think it's served me well.'

'To put the record straight, I hope!' Jim interjected, unable to let it go.

Helen met Tim several times. Before he left for the ISS, she gave him a copy of the book she herself had taken into space, the English translation of Yuri Gagarin's account of his pioneering voyage. 'From the first Brit in space to the second', she wrote inside and wished Tim well. A gesture of goodwill and a gentle reminder, perhaps, of who got there first.

'You must have been very young at the time, but do you remember Neil Armstrong landing on the Moon?' Jim asked.

It was hard to escape. Endless activities were organised at school to celebrate the Moon landings. 'When you are six everything is exciting, isn't it?' she said. 'Caterpillars turn into butterflies. Oh my goodness!' As far as six-year-old Helen was concerned, a man walking on the Moon couldn't compete with metamorphosis and countless other everyday but nonetheless extraordinary events. Often during the summer of 1969 Helen found herself thinking, 'Oh no, not the Moon again!'

Her father, a physicist, 'clearly very, very frustrated' by the fact that she 'didn't get how important this was', dragged Helen off down the garden. 'There's the Moon! There are people now walking on the surface of the Moon!' he said, desperate to impress. And Helen remembers thinking, 'What's so great about walking? I learnt how to walk years ago.'

Having been the only girl in her class to do science A

Levels, she read chemistry at university. She probably preferred physics but 'couldn't possibly do exactly the same thing' as her father, and went on to work for the General Electric Company in Coventry. From there she moved to London to work for Mars Confectionery, turning a chocolate bar into an ice cream.

'I was really interested in freezing stuff that had sugar dissolved in it and watching the melting point gradually decrease,' she said. The higher the sugar content, the lower the melting point, and so the challenge was: how much sugar could you add to a Mars bar before it would no longer stay frozen in a domestic freezer?

Once the ice-cream Mars bars were in production, she moved to the chocolate department. 'I love operations. I love seeing the way things process from one point to another, especially when you've got a physical product at the end . . . It's wonderful.'

'Even better when you can eat it,' Jim suggested, smiling.

Driving home from work one day, resigned to the traffic on the M4, she was flicking through radio stations searching for some 'decent music', when she heard a rather unexpected radio advertisement: 'Astronaut wanted. No experience necessary', it said.

'Was it just at that moment that you suddenly thought, "This is what I want to do"?' Jim asked, wondering if it really was a snap decision to apply or if she had always quietly dreamt of being an astronaut.

'I had never, ever imagined that I was going to go to space,' she said. 'British people didn't do that kind of thing.' Nonetheless, she made an effort to remember the advertised

'I had never, ever imagined that I was going to go to space'

number and later made a phone call. 'I answered a few questions over the phone and was sent an application form with that awful blank page at the end.'

'To be honest, I wasn't so much thinking about space flight itself,' she said. But the idea of doing experiments in space appealed to her. And she had other reasons for applying. She liked the idea of learning Russian, plus an intense programme of physical training sounded like an excellent way of getting fit. Training to be an astronaut: a work-out for the body and the mind. The application form, completed one evening after work, was 'a shot in the dark. I honestly never expected a response.' She now wishes that she had made a photocopy of what she wrote.

Being called for the first round of medical and psychological tests was a big surprise. 'Was it tough?' Jim asked.

'No. Perhaps that's why I was chosen in the end,' Helen said. The Soviets knew what they were looking for. They had been training cosmonauts since the 1950s. Yuri Gagarin was the first of many. An early test, which Helen aced, involved assessing how much acceleration due to gravity the wannabe astronauts could withstand without passing out. 'It wasn't a problem for me to tense my muscles and keep the blood pressure high in my brain so I wouldn't black out.'

More space-flight-specific challenges followed, revealing further hidden talents. Helen didn't mind being spun around while moving backwards and forwards or spinning while moving her head up and down.

'Do a lot of candidates puke their guts out?' Jim asked, cutting to the chase.

'Either that or they feel so bad they have to stop,' Helen said. 'Or they couldn't manage afterwards.' A graph plotting everyone's bodily responses to this challenge was revealing. 'It was quite odd,' Helen said. 'Most people are in the middle somewhere, coping more or less well. And then there was an outlier, and that outlier was me.'

Did Helen's decidedly un-queasy constitution raise her above the rest? 'It was about being a normal kind of person, both physically and mentally,' Helen insisted, refusing to recognise her exceptional, if rather niche, ability to spin around in multiple different ways simultaneously without being sick. A fairground waltzer plus, plus, plus.

Of the 13,000 young Britons who replied to the radio ad, four were selected to train in the Soviet Union, live on TV in November 1989, an experience she dreaded. 'It was very cringe-making as far as I was concerned,' she said.

Helen's parents were in the audience. How did they react? 'I don't remember anything in particular. It was probably just a "Great, well done!" Not letting me get too big for my boots, I'm sure.'

'Still, they must have been pretty surprised. Proud? Worried?' Jim added, amazed by such parental nonchalance.

'They never showed any sign of worry. Which was rather nice.'

'Were you surprised that you had got that far?' Jim asked.

'I was surprised at every single step of the way,' said Helen.

Of the four who made it to the final, only Helen and Timothy Mace were selected to undergo full-time training in Star City, a military camp 30km outside Moscow. The other two (Clive Smith, a university lecturer, and Gordon Brookes, a navy doctor) were perhaps a little disappointed to discover that they were not going to be astronauts after all. Their job would be to train the others to do experiments on *Mir*, without any chance of going there themselves.

'On paper at least, you and Timothy Mace couldn't be more different . . .' Jim said.

'Those differences were actually very useful,' Helen said. 'Especially in the beginning when the Russian language was quite tough.' . Helen helped Tim with his Russian homework and understanding the science experiments. And Tim helped her with lessons on ballistics and flight. 'So we worked together quite well'.

'But you both knew that ultimately only one of you would be going on the mission and the other would be the backup, the understudy?' Jim said.

'It was quite a weird relationship in that respect. There was an element of competition and yet we needed each other.'

They relied heavily on each other 'to get through some of the frustrations of the day'. It wasn't always easy to cope with 'not being as independent as we had been used to being in the UK'.[1] A mutually dependent couple competing for a single seat in a *Soyuz* capsule.

After a few months, the training became more practical. They spent a lot of time in simulators, experiencing what space would be like without leaving Star City. 'Everybody loves the weightlessness training,' Helen said.

'It's like being on the steep falling bit of a rollercoaster, permanently,' Jim said, perplexed.

'There's a series of parabolic loops because we're falling inside this falling aircraft,' Helen said, waxing lyrical.

---

1 Helen's visa was valid for Star City and Moscow only, and she was actively discouraged from leaving the military camp. The Star City doctors were worried she was going to get mugged or, just as bad, 'catch a cold' or some other infectious disease, but she loved her excursions on the electric train and continued to escape regularly, regardless, and was repeatedly stopped by local police and questioned.

'What a wonderful feeling!'

'It's horrific,' Jim replied, reminding Helen that it's called the 'vomit comet' for a reason.

'Most people get on fine when they're weightless for a prolonged period,' Helen continued. It can take a few days but they normally get over the motion-sickness feeling. But even the best training and the most careful selection process is no guarantee that astronauts will have a vomit-free journey. 'There's not much we can do about that,' Helen said, matter of fact. The right stuff, it seems, can't always control his or her guts.

After 18 months of training, the launch date finally arrived. And Timothy was told he would not be needed. Jim asked Helen why she thought she was selected for the mission and not Tim.

'Hard to tell, really. He would have done a great job. He passed all his exams. He was really great in space,' she said, presumably referring to a simulated universe, not the real thing. 'He would have flown instead of me if I'd have got a cold on the day of the launch.'

Eventually Jim pushed Helen to think about why she might have had the edge over Timothy. 'Maybe it was team-work,' Helen suggested. 'I got on very well with my crew.'

In the early days of space travel, it was jet-fighter pilots who became astronauts. Timothy was an Army Air Corps pilot with a degree in aeronautical engineering who went parachuting for fun and had competed in skydiving world championships. Given the choice between this daredevil *Boy's Own* hero and Helen, the girl who worked in a sweet factory, it seemed clear in many people's minds who would make the better astronaut. 'I think everyone assumed it would be him, not the girl who made chocolates,' Helen said.

'There is this idea that astronauts are almost superhuman, isn't there?' Jim said. 'In the early days, it was jet-fighter pilots who became NASA astronauts . . .'

'We need quite different people nowadays. Astronauts need to know not only when to take the lead but also when to be led'

'We need quite different people nowadays. Astronauts need to know not only when to take the lead but also when to be led. You want people who will collaborate well. Who will share their ideas when appropriate and work as part of the team.'

'And that's the kind of person you are,' Jim said, trying to get Helen to sing her own praises.

'I hope so,' Helen replied. 'I'd like to think so.'

On 18 May 1991, Helen and two Soviet cosmonauts, commander Anatoly Artsebarsky and flight engineer Sergie Krikalyov, strapped themselves into the *Soyuz* TM-12 capsule destined for the Soviet space station, *Mir*. Five years earlier, NASA's Space Shuttle *Challenger* had exploded a minute after its launch, killing all the astronauts on board – a disaster that was broadcast live on TV.

'Was it tremendously exciting?' Jim asked. 'Or were you somewhat scared?'

'At the end of 18 months, all you want to do is to stop talking and get on with it . . . It's such a relief!' Having practised every possible manoeuvre hundreds of times, an end to the monotony was welcome. 'I don't believe anyone is really scared on the day of the launch . . . It's the unknown people are scared of. And after you've been through all that training, you don't feel there's any unknown left.'

As the *Soyuz* capsule soared into space, Helen sat inside 'feeling hot and gradually more uncomfortable ... strapped into the same seat, in the same position, in the same space suit ... sweating buckets.' There were so many procedures to follow, she nearly forgot to look out of the window. When she did, her eyes slowly got used to the darkness. As her pupils dilated, stars started to appear. 'The more you look, the more stars you can see. It really does feel as though they go on for ever.' Looking backwards, she could see the curvature of the Earth.

On arrival at *Mir* five hours later they were given elastic straps to tie tightly around the tops of their legs so as to relieve the venous pressure that results from zero gravity. Doing this made Helen's upper body feel 'a bit less uncomfortable' but her legs felt worse. 'So, I rapidly dispensed with them and just put up with this stuffy feeling,' she said. After a few days, however, things improved. The brain realises that there's too much fluid around and tells the kidneys to excrete extra urine, Helen explained. Typically, two litres more than normal.

Helen mentioning urine gave Jim the opportunity to ask the question that Helen is asked more often than any other: How do you go to the loo in space?

Being in a spacesuit was the least of her problems. Without gravity to direct things, excretion has the potential to be very messy. 'We flush our toilets with air rather than water,' Helen said. So far so good, but further explanation made the whole business sound rather less dignified. Astronauts do what they need to do, locating the relevant body parts as close as possible to a funnel. Solids and liquids in freefall are then hoovered up, pulled through a tube and stored, ready to be recycled.

'Psychologists think that it's not very good to drink your

own toilet water.' Helen, however, would have been happy to drink recycled urine. It was 'no big deal'. 'Of course, scientists know that it's what the world has been doing for aeons,' she said. 'With some heat-exchange processes and some reverse osmosis and finally with silver ions, we can actually create water that's clean enough to drink.'

She drank the water prescribed by the psychologists reluctantly. What was the point of taking fresh water supplies into space when there was perfectly good water made from urine on board? It was, she thought, a ridiculous waste of *Soyuz* TM-12's precious payload. At least they were allowed to use recycled urine in place of water for their experiments. And recycled urine was used in the air conditioning. Oxygen was extracted from the water in urine by passing an electric current through it and then used to oxygenate the cabin air, which otherwise would become too rich in carbon dioxide as the astronauts exhaled. 'So you can actually breathe what you put in the toilet rather than drinking it. For some reason, the psychologists think that's a much better deal.'

There was not a lot of freedom in space. Helen's time was controlled 'almost minute by minute' by mission control. Timers were set for all the different experiments. She was told precisely when to wake up and when to go to sleep. Most of the time she didn't mind. But 'we didn't stick to everything' she said, a little sheepish. 'We didn't always go to sleep when we were told to go to sleep.' There were not many opportunities to exercise autonomy, it seems.

'You're probably not supposed to admit to this,' Jim said, 'but did you ever feel a bit unsettled being so detached from mother Earth? Or were you always completely at ease?'

'I felt fine,' Helen said. 'And I think it's because I wasn't on my own.' Yuri Gagarin flew solo. 'That would have been quite different,' Helen said.

She enjoyed the 'slight detachment it gives you. I liked being able to look back at planet Earth and look back at my life.' 'I like climbing mountains for the same reasons,' Helen said, bringing things back down to Earth.

'Was it a very bonding experience for the five of you up there, so separate from the rest of humanity?'

'The bonding experiences, I think, come from the tough times, whatever in life you're doing,' Helen said. One of the toughest moments on this mission was when they approached the *Mir* space station and the automatic docking system failed.

All those simulated landings in Star City reduced the stress. Reality was easily mistaken for yet another practice drill. 'The only difference this time was that if we failed, this really was our lives we were putting at stake.' Not insignificant! They had enough fuel to orbit the Earth one more time, but if the second docking failed, they had to return to Earth. 'So that's your space mission over,' Helen said. 'If you missed by a small distance, you could crash into the space station, damaging it and the spacecraft, and maybe not coming back alive . . . You're relying on each other, trusting each other with your lives.'

After eight days in space, Helen returned to Earth. Jim asked if she would have liked to stay in space longer.

'Three months would have been ideal,' Helen said. 'It's so much effort getting into space, it makes sense to spend as long there as you sensibly can.' But it would 'get a bit tough' after a while, she thought. She would miss friends and family and all the people she meets commuting and

'It's so much effort getting into space, it makes sense to spend as long there as you sensibly can'

on shopping trips. 'Even if you don't talk to them, you get input from them all the time,' she said. 'I would miss all that human interaction.'

If being on TV for the first time (when she was selected for the mission) was cringe-making, the media attention Helen received when she came back down to Earth must have been excruciating. While she was in space, 'the 27-year-old from Sheffield' had become a national hero.

'Did you feel that you somehow needed to be the person everyone wanted you to be, to project this heroic image of yourself, regardless of how you might be feeling inside?' Jim asked.

'It is rather strange seeing yourself in the newspaper,' Helen replied. Being in the public eye was clearly not something she enjoyed. 'I've never talked about my private life and I think that's how I've dealt with it.'

'It must have made it difficult for you when the media would keep on asking you personal questions?' Jim said.

'Oh, I just say no,' Helen said, abruptly. 'That is the only way to do it . . . I was never tempted because I just know absolutely where my boundaries lie.' For eight years she travelled up and down the UK, happily giving talks about space. To this day, she remains silent about even the most simple questions about her personal life. 'I didn't go into space to be famous,' she said. 'I'm a normal kind of person and if that is what the British public get out of it, well, that's great.' Normal kind of people go into space. Normal kind of people can do amazing things.'

BRIAN COX

'Why shouldn't scientists be seen or spoken about in the same breath as footballers and people on The X Factor*?'*

Grew up in: Oldham
Home life: married to TV presenter Gia Milinovich with two children
Occupation: physicist
Job title: Professor of Particle Physics, University of Manchester
Might have been: a pop star
Inspiration: Carl Sagan
Passion: quantum mechanics
Mission: to make science part of popular culture
Date of broadcast: 23 September 2014

A physicist who is regularly snapped by the paparazzi, Brian Cox is the UK's first science celebrity – the pin-up of particle physics. Voted the 'sexiest man alive' in a survey by *Time* magazine in 2009, he insists nonetheless that he is 'a simple-minded northern bloke'. His first major BBC TV series attracted six times more viewers than were expected, and fans flock to *The Infinite Monkey Cage*, which he hosts for BBC Radio 4 with comedian Robin Ince. The 'Cox effect' is thought to be responsible for a boost in university admissions to read physics. Brian worked at CERN when the Large Hadron Collider was being built and spends a lot of time thinking about the behaviour of those exotically named elementary particles: quarks, neutrinos, gluons and muons. For all his celebrity status, he continues to publish papers on quantum mechanics and is convinced that the existence of many universes makes more sense than there being just one.

At home in Oldham on Saturday nights, the young Brian Cox would watch *Tomorrow's World* and *Top of the Pops*. 'That pretty much sums me up, I suppose,' he told Jim.

After leaving school, he played keyboards with the rock band Dare and toured with them for five years, before deciding that studying subatomic particles might be more worthwhile. When he appeared on *Top of the Pops* in 1994 performing 'Things Can Only Get Better' with D:Ream, it was a childhood dream come true.

'So, Brian,' said Jim. 'Have you been taken aback by all this adulation?'

'I'm not surprised that people get excited by the ideas in particle physics,' Brian replied, failing to acknowledge (perhaps deliberately) that some might not. That physics is endlessly fascinating is a given for him. 'But it does surprise you,' he said. 'And it surprises me that I have been the one who has been most subject to that kind of attention.'

'Attention on you in particular,' Jim said. 'I mean, we're talking screaming teenagers. You can't travel on the London underground without being mobbed. And this is all for someone who spends his time trying to understand the nature of quarks, quasars and quantum mechanics. It is remarkable.'

'It is,' Brian replied, calmly. 'Popular culture is a fickle thing, but why shouldn't scientists be seen or spoken about in the same breath as footballers and people on *The X Factor*? We're more important than footballers or pop stars.'

He 'always thought science was too important not to

be part of popular culture' but was expecting evolution, not revolution. He didn't think he would see it happen in his lifetime.

Brian left school all set to study electronic engineering at Leeds University, 'as advised by the careers people', having spent a lot of time as a teenager building computers. (Inspired by the new sounds that were synthesised by Orchestral Manoeuvres in the Dark, Duran Duran and Ultravox, he wanted to do the same.) But university was put on hold when the drummer from Thin Lizzy moved in down the road. Brian's dad had met him down the pub when he was on the rebound after Thin Lizzy had broken up. Brian became the keyboard player for his new group, Dare, which turned into a 'biggish band', recording three albums in Los Angeles.

'So, I just ended up taking five years off accidentally, really,' Brian said.

'Was there a make-or-break moment that took Brian Cox the keyboard player to Brian Cox the particle physicist?' Jim asked.

'There was,' Brian said. 'Life is always slightly random and can turn on the smallest of events. The band had just had a fight, that was the trigger,' but it had been building for a while. Brian was 'getting a bit bored' with being a rock star. 'I do get bored very quickly,' he continued, then clarified (perhaps just in case anyone should think he was suggesting physics was boring), 'but I never get bored with physics.'

'Life is always slightly random and can turn on the smallest of events'

As a particle physicist, he is

interested in what happens inside atoms. The Standard Model of particle physics tells us that there are two families of fundamental particles: quarks and leptons. These are particles that we have mathematical (as well as practical) reasons to believe cannot be broken down any further. Quarks come in six different flavours: up, down, strange, charm, bottom and top. Leptons include electrons, muons and taus, and three types of neutrino. And gauge bosons carry forces. They include photons (which bring light) and Higgs bosons (which give things mass).

Welcome to the wonderful world of quantum mechanics, often also described as weird. To Brian, it is wonderfully simple, if only we could let go of the old ways of thinking. We just need to stop imagining that subatomic particles behave like mini-versions of bigger things. They don't. Newton's laws, described in 1686, have served us well. We can accurately predict where apples, mechanical parts and planets will be. We can send rockets to the Moon, Mars and the distant comet 67P. But inside the atom, these laws of motion do not apply. It is a quantum world and the mechanics are quantum too.

'From Newton onwards, we knew that if we could just calculate everything with sufficient precision, we can tell you what's going to happen,' Brian said. 'Quantum mechanics is subtly different.'

Most people generally like to know where things are (uncertainty is unsettling), but inside atoms probabilities are as good as it gets. Ask a particle physicist about the whereabouts of a quark and they might say something like, '70 per cent here, 27 per cent there and 3 per cent over there.' This is not a fudge. Nor are the physicists hedging their bets. A suite of possibilities is described because, inside the atom, uncertainty rules.

The randomness of radioactive decay gave us the first clue. Werner Heisenberg sealed the deal. His Uncertainty Principle (which states that you can know either the position or the momentum of a subatomic particle, but never both) shocked everyone. Even Einstein was unhappy. 'God does not play dice,' he pronounced. But as the flamboyant, brilliant and much-admired (by Brian and millions of others) Richard Feynman said, subatomic particles 'do not behave like waves, they do not behave like particles, they do not behave like clouds or billiard balls, or weights and springs, or anything that you have ever seen.' And the laws that govern them are unfamiliar too. They are described, not by Isaac Newton, but by Max Born.

The Born rule describes the probabilistic nature of our universe at its most fundamental level. It allows us to calculate the possibility that a particle will be in a particular place (based on measurements of time, mass and distance).

'And you can compare that with experiments, and it works.' Phenomenally well. Brian once claimed that the rules of quantum mechanics can be written on the back of an envelope, so Jim invited him to do something similar.

'So, Brian Cox, without hesitation, repetition or deviation, can you explain, as succinctly as possible, the rules of quantum mechanics? Your time starts now'

'Well, the most basic version I know of it is Feynman's version, which essentially says particles are particles and they hop from place to place with a particular probability. And the probability that a particle that's at one place will be at a different place later is given by a very simple rule. It uses a quantity called the action which is to do with the mass of the particle, and the time and the distance. So, you basically calculate these little quantities – you add them up.

'So, I start with an electron in one corner of the room,

and I say, "What's the probability it will be somewhere else in the room later?" Then at every point in the room, you can assign a probability that it will be there at a later point, with one simple rule. And that's it.

'And this is called the path integral formulation of quantum mechanics. And it underlies everything else. You can get the rest from that. So, it's a simple rule. What's the probability of a particle moving from A to B? That's it.'

If we can embrace Born's rule (and let Newton go), there's no need for the mind to be boggled by quantum physics. What the rules of quantum mechanics tell us is simply this: there are lots of possibilities, some more likely than others. No one could call that weird.

What's more, these probabilities can be predicted with absolute precision and then tested in the real world. This wasn't possible when Born came up with his rule, but it is now, thanks to the Large Hadron Collider and other particle accelerators around the world.

To explore inside the nucleus of an atom, first it needs to be broken down into its constituent parts: protons and neutrons. Then protons or neutrons can be smashed together, liberating quarks. In a bagel-shaped tunnel buried beneath the Swiss–French border, the most energetic collisions ever achieved on Earth have taken place. Protons have been accelerated to just under the speed of light, travelling in opposite directions around a tunnel that, at 27km, is as long as the Circle Line on the London underground and has similar proportions. When collisions take place inside the Large Hadron Collider, huge digital cameras record what happens, and many other measurements are made.

During and after his PhD, Brian spent many years working at CERN in the 1990s when the LHC was being built.

'I had a great time working with the accelerator engineers at CERN,' he said. 'They're terrific!' And he enjoyed working with the scientists at the LHC, learning about accelerator physics. He was writing computer programs to describe what the Standard Model would expect to happen when the high-energy collisions that were planned at the LHC actually took place.

On 10 September 2008 the LHC was turned on, timed to coincide with the *Today* programme on Radio 4. When BBC Radio Science Unit producer Alexandra Feachem had asked politely if this would be possible, a roomful of CERN physicists were, at best, bemused. Some were downright rude. But Brian helped. If physicists don't share their work, why should the rest of us care? And future funding for the LHC, he reminded them, was far from guaranteed. Together he and the BBC Radio Science Unit triumphed.

Switch-on was broadcast live, with Brian in the control room surrounded by journalists, all primed to celebrate 'Big Bang Day'. When the first proton beam completed a lap of the LHC, it was global news. Some people feared that such a high-energy collision could generate an all-consuming black hole. The story for them was that the world might be about to come to an end. But no matter. Millions of people now knew about the Large Hadron Collider. Particle physics had entered popular culture.

Nine days later, there was a major problem. Before the first collision had even taken place, there had been a serious explosion, severe enough to rip magnets out of the concrete that was supposed to hold them in place. But a year on, the problem was fixed, and the results started rolling in. Thousands of scientists across Europe (Brian included) were finally able to see how their computer-generated data (as predicted by the laws of quantum mechanics) matched up with the real world, in the process finding out if the theory of quantum mechanics accurately described subatomic reality. Many had waited many years for this moment.

'It was one of the most remarkable feelings I've ever had,' Brian said. 'You type in all these things, all your theory. You make predictions and you look at these collisions. And they just match in every detail. They match the theory.' Joy! This was the experience, Brian told Jim, that made him understand the power of science.

'You type in all these things, all your theory. You make predictions and you look at these collisions. And they just match in every detail. They match the theory'

All good science experiments are written up in the same way: method, results, conclusion. This is what we know so far about the Large Hadron Collider experiment. The method (colliding hadrons) works. Smashing protons together in particle accelerators has liberated quarks and other subatomic particles, and we have found ways to measure how many of these exotically named particles behave. There is no shortage of results. Now all that's needed is a conclusion. And on this, physicists disagree. Often quite profoundly.

Brian and his good friend and colleague Jeff Forshaw have worked together for many years and have written several books together about what happens inside atoms.

'We're almost exactly the same age, and we were born geographically roughly in the same place,' Brian said. And, perhaps most important of all, 'We think in the same way about physics.'

Despite all their similarities, even they don't always see eye to eye about what it all means. 'We were working on one of our science books,' Jeff said, 'trying to get our head around understanding quantum mechanics, and we spent until the very early hours arguing, really shouting at each other.' Why? Because they couldn't agree on what might seem to be a rather important point: is there just one universe or are there many? Or to put it more technically: should we adopt the Copenhagen interpretation of quantum mechanics or Everett's 'Many Worlds' idea?

'It was one of those situations when we were convinced the other person was talking rubbish,' Jeff said. According to him, one chapter of their book *The Quantum Universe* was 'very heavily influenced by that drunken rant'.

'We do like getting drunk and arguing about physics,' Brian confirmed. 'And we'll do it all night!'

The Copenhagen interpretation of quantum mechanics states that the cat in Erwin Schrödinger's 'cat in a box' thought experiment is both dead and alive at the same time, until someone opens the box to check its fate, at which point the system will decide: is it dead or alive? In the Many Worlds interpretation, there are many universes. There's a universe where the cat is dead and another separate universe in which the cat is alive, and many more besides in which the cat exists in differing ratios of simultaneous deadness and aliveness.

'What side were you on?' Jim asked Brian. 'Copenhagen or the Many Worlds interpretation?'

'No,' Brian said firmly. 'I don't like the Copenhagen interpretation. Saying there's an infinite number of universes sounds more complicated than there being just one, but, actually, it's a simpler version of quantum mechanics. It's quantum mechanics without something called wave function collapse, which is the idea that by observing something, you force a system to make a choice: is the cat alive or dead? Quantum mechanics might tell us, for example, a probability that the cat is 75 per cent alive and 25 per cent dead. Everyone agrees with that. That's how the theory works.'

The Copenhagen interpretation is based on the assumption that the cat must be either dead or alive. The Many Worlds interpretation thinks this is a mistake. Instead, it simply accepts that the cat is both 75 per cent alive and 25 per cent dead. 'And it leaves it at that,' Brian said. 'The thing has both these possibilities in it, in a mathematical sense.'

And the problem then becomes, 'How can it be that something like the cat, or me, can be 75 per cent alive and 25 per cent dead? That's where the interpretation comes in, but it's a simple basic proposition.'

'So, you're saying, despite the proponents of Many Worlds arguing that there are parallel universes, which tends to scare people, including physicists, it's actually a simpler, more logical way of getting to grips with the counter-intuitive nature of quantum theory?'

'It is,' Brian said, concisely. He believes our desire for definitive answers can prejudice our understanding. We lack the humility to accept that our perceptions are crude. Why can't the cat exist as an ensemble of possibilities? Reality might be multi-layered, rather like an image in Photoshop. For example, 75 per cent alive and 25 per cent dead; 80 per cent dead, 20 per cent alive; 0 per cent alive and 100 per cent dead. 'So, you can imagine layers where all these different possibilities play themselves out,' Brian said. 'We call them Feynman diagrams in particle physics. And the Many Worlds interpretation just says, "Well, that's how reality is. All these layers, all these possibilities, are still there."'

Plenty of people are now working on trying reconcile these two views. 'But I get the sense that more and more physicists are tending towards the Many Worlds interpretation,' Brian said. 'I really get that sense, which is very interesting.'

As Brian spends so much of his time inhabiting multiple universes in his mind, he can perhaps be forgiven for being a little otherworldly himself. According to his friend, the comedian Robin Ince, 'he floats through the world, unaware of a lot of the other things that are around him.'

'When I'm crossing the road with him,' Robin said, 'I am almost as aware of his footfalls as I am of a child's. You have to hold his elbow to make sure that halfway across, as the number 37 bus speeds down the road, he doesn't suddenly think of something magnificent about pulsars, stop for a moment, look at the sky and find himself under the wheels.'

'That familiar image of you on TV, gazing up at the stars in all amazement, at the sheer beauty of it all – is that a pose that you've cultivated for TV or is that what you're really like?' Jim asked.

Brian ducked the question. Presenting TV programmes helps him to rediscover the sense of wonder that he enjoyed when he was five years old and looked at the sky and thought, 'There's something about astronomy I like.' Developing a better understanding of his own motivations is, he said, 'one of the great pleasures and privileges of making television'.

Brian's TV career began with another 'small and slightly random event'. Lord Sainsbury had resigned from being science minister in 2006, and 'there was a series of jobbing science ministers who didn't pay attention'. Mourning the loss of Lord Sainsbury, 'a great science minister', he started getting involved with the media, 'for political reasons in some sense'. And he ended up on television quite a lot, 'whining and whinging and shouting on *Newsnight*! And that deflected me off into media and political engagement.'

As his media career took off, the Royal Society might have grumbled. He had been awarded a highly prized ten-year university research fellowships a few years before, in 2005. Instead, 'to their absolute credit', they said, 'It's fine.'

'They didn't come back to you and say, "That's not what we're paying your salary for. Go back to CERN!"?' Jim asked.

'They really didn't. They gave me freedom to develop into a scientific leader. That was the language they used.' If Brian was engaging with the public and politicians, and doing research, then that was good. 'And I think that's important and to be applauded,' Brian said. 'That's important.'

'You really do seem to want to inspire people with science. Why does that matter to you so much?' Jim asked.

'It matters because I think that our civilisation is built on science, in the widest possible definition. I genuinely believe that our economy is a knowledge-based economy. It's built on education, it's built on research, science. And also on research into social sciences and the arts. It's built on intellect. And countries like ours, which I want to do well, because I love it, will do well by investing in knowledge.'

Introducing people to science, Brian believes, is to give them access not only to joy and wonder, but also 'actually to riches'. 'Knowledge is wonderful in itself and exciting and beautiful. It also grows economies.' And if professors engage with the wider world, universities, 'those tremendously powerful organisations sat in the heart of cities' where intellectual debates take place, can be engines of social change. 'And that's a great success, I think, over the last ten years.'

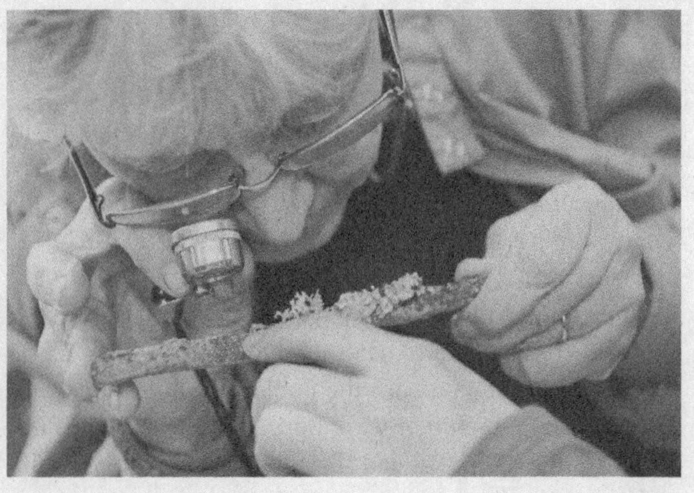

OPAL www.opalexplorenature.org

PAT WOLSELEY

'There are almost as many different species of lichen, as there are species of flowering plants'

Grew up in: Cheshire, Aden and North Wales
Home life: two sons, Will and Tom
Occupation: lichenologist
Job title: Science Associate, the Natural History Museum, London
Might have been: an artist
Inspiration: Tŷ Canol, an ancient Welsh woodland
Passion: lichen
Mission: to understand how lichen respond to changing environments
Best moment: realising the crustose lichen in Tŷ Canol must be thousands of years old
Advice to young scientists: just keep going
Date of broadcast: 14 August 2012

Pat Wolseley fell in love with lichen when she attended a short course run by a lichenologist. Her eyes were opened to the beauty and incredible diversity of these unassuming organisms, stealing her attention away from water plants. She started her lichen career identifying hundreds of different species of lichen in an ancient Welsh woodland and went on to study lichen as indicators of environmental change, using them to measure humidity levels in tropical rain forests and monitor levels of atmospheric ammonia in farming areas in Britain. Most recently, she developed low-tech ways of using lichen to monitor air pollution in our cities, and engaged hundreds of volunteers in Open Air Laboratories, a nationwide project to encourage as many people as possible to study nature and collect data.

Little loved and often overlooked, lichen cover nearly one-tenth of the Earth's land surface. They grow slowly and are fantastically long-lived, as if everything about them is slowed down and stretched out, and they can be found on trees, roofs, rocky outcrops and tombstones:[1] almost any surface will do. *Lecanora muralis* sticks like chewing gum to city pavements. Methuselah's beard hangs from conifers, growing up to three metres long.

'Once you start looking, lichen are everywhere,' Pat reminded Jim, perhaps hoping to entice him in. Some species are flat, others bushy; some powdery, others more like jelly. After the *Braer* oil-tanker disaster off the coast of the Shetland Islands in 1993, tar spot lichen was mistaken for splodges of oil on the rocks. 'There are almost as many different lichen as there are species of flowering plants,' Pat said.

Perhaps detecting a bit of chippiness, Jim asked: 'Is it a bit of a Cinderella science?'

'Undoubtedly, undoubtedly,' Pat replied with gusto. 'So we have to push it up.'

'What is it about lichen?' Jim asked. 'Was their beauty part of the attraction for you?'

'Yes. Definitely. Definitely,' Pat replied. 'There's this incredible variation in pattern and colour, even in a single branch . . . It's like having a third eye,' Pat said, suddenly dreamy. And then she added, perhaps realising that not everyone would see lichen in the same way that she did: 'You have to get hooked, mind.'

1 One-third of all British lichens are found in graveyards.

She examines different species of lichen, looking for subtle variations in colour, shape and form, exploring the world with an artist's eye. 'It's amazing what you can see just by looking.'

Pat went to boarding school when she was five and sat the Oxford entrance exam at a wonky table in the headmistress's office at Plymouth Brothers School in North Wales. She studied botany at Oxford University when only a few colleges admitted women, completed an art foundation course at Camberwell College of Arts and did an exquisitely illustrated research project on water plants. 'I am still hugely interested in art,' she said, 'but I like the questions in science.' Questions that are never answered and only raise more questions, 'So you never quite finish,' Pat said, evidently thrilled. 'And I suppose, we scientists. We don't stop!'

'I am still hugely interested in art but I like the questions in science'

It was the lichenologist Peter James who provided lichen bait for Pat, when she attended a short course at the Nettlecombe Field Studies Centre in Somerset, 'highly pregnant' with her first child. 'It took a while' before she found a way to channel the enthusiasm for lichen that Peter had inspired. But several years later, when Peter was looking for people to do fieldwork in an ancient woodland in West Wales, Pat volunteered. By now she was a single mother of two young boys, having been married to an artist for many years.

When she started working in Tŷ Canol, 143 different species of lichen had been recorded in the 38-acre wood.

Pat identified 400 varieties in three months. 'It made me think, "What else is out there?"' she said.

She has spent the rest of her career trying to find out, doing fieldwork in the UK, Iceland, Sri Lanka, Borneo and Thailand. A project to study lichen in Huai Kha Khaeng, a wildlife sanctuary in a remote region of Thailand, near the border with Burma that was designed to last for five years (1989–1994), is still going strong. On a recent visit Pat saw a park ranger enthusiastically sharing his newly acquired knowledge of all the different lichens with all the other rangers. His proselytising, she thought, would do more to make people in Thailand aware of the wonderful diversity of lichens on their doorstep than any scientific paper.

The research plan was to use lichen as indicators of environmental change. People were starting fires to clear the ground and promote the growth of valuable fungi which thrive on charred remains, and there was concern that these fires were drying out the rainforest. Some species of lichen love moist conditions and others thrive when it is dry, so by studying their relative abundance over time, Pat hoped to gain an insight into whether or not humidity levels in the forest had changed.

Working with several Thai colleagues, she created a permanent plot, deep in the rainforest, and identified dozens of lichen sites within it. 'We wanted to be able to go back to the same tree, the same site and see what had happened,' Pat said. She drew a beautiful map and they put up signs (hoping they wouldn't get swallowed by the forest over time), marking branches that were of interest with a dab of paint, often climbing trees using ladders that they had made from the bamboos growing nearby.

She collected lichen samples with a custom-made

machete. (Some tropical barks are very hard indeed; a penknife just doesn't cut it.) Once when Pat was cutting lichen off a tree, her machete slipped and she nearly cut her finger off instead. According to her colleague, who 'nearly collapsed' on seeing her dangling digit, Pat 'was very calm about it'. She promptly tore a strip off her shirt, wrapped it around her finger to hold it together, and moved on.

'And did you discover many new species?' Jim asked.

'We couldn't always identify them,' Pat replied. 'You didn't necessarily know if you had a new species.'

'So it might be new to you . . . but it may well have been studied and catalogued by someone else?'

'Absolutely.'

'And how did you record what you saw yourself? Were you hacking off samples and sticking them in your rucksack to take back?' Jim asked.

'Yes, they do have to go back. Hopefully not in one's rucksack. But the most important thing is that those samples are preserved and catalogued so that we can then work on them and describe new species, and so they can be made available in a collection somewhere. The most important thing for our knowledge of tropical biodiversity is that somewhere on this Earth, there are good collections that are named.'

'The most important thing for our knowledge of tropical biodiversity is that somewhere on this Earth, there are good collections that are named'

Days were spent gathering specimens and sticking then onto pieces of white card, then writing detailed notes on the back. 'You can have a jolly good guess at what species you've got in your hand,' Pat said. Some species, nonetheless, can only be confirmed under a microscope or by doing a chemical test. Back in London, she consulted established lichen collections to see if what she had found was new just to her or new to everyone. 'It's easy to forget that other people may have been there before you, often in a different century,' she said, smiling.

'What about modern technology?' Jim asked. 'Can that help?

'It's a brilliant way to go!' Pat said, excited – iPads, not index cards, may well be the way forward. High-resolution cameras mean samples can be observed in incredible detail and the image preserved and shared. Pat has developed an iPad key to help fieldworkers to identify different species by encouraging them to think systematically about their different characteristics. And the details of every lichen currently known to humankind can be stored and recalled. The most powerful computer in the world, however, will only ever be as good as the data it has been fed. (Garbage in, garbage out, or GIGO, as the computer scientists like to say.) And Pat believes there will always be a role for good old-fashioned taxonomy.

Lichen, once classified as plants, are now classed as a special group of fungi: they are fungi that live in symbiosis with an alga or a cyano-bacteria. The fungus provides its single-celled friend with a comfortable, well-lit home, in return for photosynthesising services that generate food. Happily, these two organisms co-exist without any need for soil. Rootless, they live on air. Like canaries in a coal mine (that become drowsy when the amount of methane in the air reaches a certain level), certain species of lichen can be used to monitor the quality of the air.

During the London smogs in the 1950s, many species were killed off by the high levels of sulphur dioxide in the air. Data provided by lichenologists was a key part of the motivation for the Clean Air Act of 1968. And after the act was introduced, making it illegal to emit dark smoke from chimneys in London, many sulphur-sensitive species returned.

By the 1990s, high levels of atmospheric ammonia had been linked to an increased risk of blue baby syndrome and the EU safe limit was under review. In response, a research project was set up to see if it was possible to monitor the amount of nitrogen in the air using lichen as indicators. 'We already knew that some lichens were highly sensitive to nitrogen,' Pat said. Farmyard roofs covered in bright-yellow lichen were a familiar sight. Certain species seemed to thrive on the nitrogen-enriched air in areas where fertilisers were generously used. Others died.

'So it's all about measuring the relative abundance of these different species?' Jim asked.

Yes. But first they needed to prove there was indeed a connection between lichen populations and atmospheric conditions. 'Of course, we couldn't get the evidence unless we actually measured the air quality as well,' Pat said. To this end, Pat and her colleagues joined forces with the Centre for Ecology and Hydrology and gained access to their data on ammonia levels. 'We had sites across the whole of Britain where we were measuring ammonia and looking at lichen communities.'

By comparing these two very different data sets, they were able to work out how the atmospheric ammonia levels and lichen communities were related. 'That's where we really developed the actual science of using this connection between ammonia and lichen,' she said. Then, using this knowledge, they were able to prove that certain nitrogen-sensitive lichens were indeed disappearing in areas with high levels of atmospheric nitrogen, while others seemed to be thriving on the nitrogen pollution. 'We were losing our bushy species and our greener species and those bright-yellow lichens were coming in.'

'And how do you learn about the history of air quality in a particular area?' Jim asked.

'That's where the lichens come into their own!' Pat exclaimed, clearly delighted with Jim's question. 'I started looking at the age of the substrate,' she said When the substrate is a tombstone with a date engraved, it's very helpful. Tree bark is less precise but useful nonetheless, Pat realised.

By far the most significant difference between the trunk of a tree and the outer branches and twigs is their age. Lichen are fussy organisms, sensitive to many things, but they are also tenacious. They grow very slowly and can survive for centuries. And so Pat concluded: 'The lichens on the trunk of a tree might have been there for a very long time (if they haven't been bumped off, as it were, by sulphur dioxide). But the lichens on the twigs and branches are telling you what it's like now.'

Such a simple idea! Why had no one thought of it before, Jim wondered.

'At the right moment, when the knowledge is there, and you've got your eye in, you suddenly see things that you haven't seen for years,' Pat said. She had spent years looking at lichen on trees without having this thought. 'There are no new ideas in science,

'There are no new ideas in science, we just put them together in a different way'

we just put them together in a different way.'

A huge part of Pat's research has been devoted to her 'twig/trunk hypothesis'. Lichen in search of light attach themselves to freshly grown bark, hoping it might take them somewhere sunny. The lichen on the trunk got established when the tree was growing up, many decades or even centuries ago. Therefore, a record of the local air quality could be found on the trunk and twigs of a single tree.

How was Pat's twig/trunk hypothesis received? 'Did people say, "Oh yes, yes, of course!"?'

'No!' Pat exclaimed, laughing. 'They said there are lots of problems! Which, of course, I knew. Lichen can get knocked off branches. You can't monitor them all the time. The twigs are out of reach . . .'

But as a rough-and-ready guide to changing levels of nitrogen in the air, it's hard to fault.

When Pat met Jim she was facing the unenviable task of interpreting data from 4,000 surveys that had been completed as part of the Open Air Laboratory (OPAL), a nationwide experiment funded by the National Lottery and run by the Natural History Museum in London. The survey was set up in August 2007 with the idea of recruiting citizen scientists throughout the UK to gather data on the environment in their area. The lichenologists (energised by Pat) had lobbied hard for lichen to be included in the survey. Data on local air quality was desperately needed, they said, at a time when air pollution had been rather forgotten and was not on the nation's mind in the same way that it is today. Lichen were a proven and highly accessible indicator: they are everywhere and can be spotted at any time of year. Many different species can be identified with relative ease. Anyone could learn to do it.

Pat created an easy identification system to be used in the survey, by adapting her own research methods and trying to make them failsafe. And she spent many months training community scientists, who were then dispatched to teach groups of citizens how to identify different kinds of lichen. They were also encouraged to promote the survey, with a particular focus on disadvantaged groups and urban areas. These areas are among the most polluted and they are also the most densely populated. This is not good for the health of people who live in towns; but it's a win–win situation for research. By recruiting an army of urban citizen scientists, data was gathered where the need for it was greatest.

'There just hasn't been enough work done in urban areas,' Pat said. Perhaps because lichenologists tend to favour more peaceful environments: woodlands and wildlife, not pedestrians and traffic lights. The inner city is not *their* natural habitat.

Forty thousand OPAL survey packs were sent out to citizen scientists and 4,000 completed surveys were returned, mostly from urban areas. As a result the survey 'received data from areas that scientists don't usually reach'. Lichenologist Peter Crittenden was in no doubt that the citizen scientists who had been involved in the OPAL survey had learnt a lot about both lichen and the scientific method, but he said it remained to be seen how useful the data would be in terms of providing hard information about pollution. Pat was optimistic. It took three months to clean up the data.

'Chucking out the nutters?' Jim suggested.

'Yes, that's right!' Pat confirmed.

After that, assessing the reliability of the data was more tricky. 'It's been a really interesting and challenging experience for me,' Pat said, 'but there's power in numbers.' When you plot all the findings on a graph it's easy to spot outliers and eliminate obvious mistakes. Statistical analysis helped to identify any weaknesses in the survey design. The results from most of the species identified were pleasingly consistent, but it became clear quite quickly that citizen scientists had struggled to identify one or two.

When Jim interviewed Pat, the results were still in the pipeline but other benefits of OPAL had already become clear. Citizen scientists, with an eye for lichen, can alert us all to changes in air quality in their towns and cities. 'It just empowers people,' Pat said. It encourages people to ask questions of their local councillors and politicians. 'We need science up front, so people can ask questions.' If the powers that be deny that there's a problem, Pat hopes people will say, 'Well look, it doesn't look good here!' Everyone will be empowered to gather evidence about the environment in which they live.

The enthusiasm for OPAL in one north London school was particularly notable. A group of teenage girls happily gave up hours and hours of their free time to study lichen, getting to know the lichen communities in parks, on pavements and at different distances from busy roads. They are all now able to identify many different species and do so enthusiastically, using these living indicators to identify the air-pollution hotspots in their neighbourhood. Perhaps the next great British lichen champion has already been hooked.

esa

MATT TAYLOR

'Plasma physics permeates everything'

Grew up in: Manor Park, east London
Home life: wife Leanne and two children, Lily and Harry
Occupation: astrophysicist
Job title: *Rosetta* mission Project Scientist, European Space Agency technical centre
Inspiration: his parents, who encouraged him to work hard
Passion: the *Rosetta* space mission
Mission: to land the robot *Philae* on a speeding comet, 4 billion miles from Earth
Best moment: 'A conference call to discuss the first findings from the mission with my children listening in'
Worst moment: 'Hearing that my dad had terminal cancer, immediately after a meeting to plan the end of *Rosetta*'
Advice to young scientists: 'Embrace failure, it's the greatest teacher'
Date of broadcast: 17 March 2015

Matt Taylor, a plasma physicist, became the Project Scientist on the *Rosetta* space mission to the distant comet 67P Churyumov-Gerasimenko in 2011 that was designed to land a robot, *Philae,* on this speeding agglomeration of cosmic dust and ice 4 billion miles from planet Earth. He joined the European Space Agency to work on the *Cluster* mission to study the space weather around planet Earth, and took charge of the science on the *Rosetta* mission after 'there had been some upheaval in the team'. Matt's enthusiasm for *Rosetta* fuelled global interest in this audacious mission, even if some people were offended by his choice of shirt on one occasion.

Jim interviewed Matt four months after the *Philae* robot lander had been released from the mothership.

For Matt Taylor, other than his wife, the *Rosetta* space mission to the distant comet 67P Churyumov-Gerasimenko is 'the sexiest thing alive'. Depositing a robot on a speeding comet 4 billion miles away from planet Earth was an unprecedented challenge. And on 12 November 2014 it looked as if the European Space Agency (ESA) had succeeded. The scientists at mission control in Darmstadt leapt out of their seats in pure excitement. A short video of one of the scientists, showing her jumping up and down and screaming, 'We did it! We did it!' went viral.

But minutes later the news came through that *Philae* had disappeared. The BBC Radio Science Unit was bombarded with questions. Everyone wanted to tweet but didn't know what to say. 'Has it landed?' they asked. 'Well, yes and no,' we replied. The scientists at mission control went quiet and looked tense. For 40 minutes everyone was in the dark. Then she reappeared. Again, everyone cheered. *Philae* had not landed, she had bounced. (Later it became clear that the harpoons and gas thrusters that were meant to ground *Philae* had failed, rendering the ice screws designed to attach her to the surface redundant.) Once again, she disappeared, before she finally settled on the surface of the comet as planned.

In the run-up to the landing, *Philae* had captured the imagination of the nation, having been described in a series of endearing cartoons as a plucky adventurer with a searchlight and a hard hat. And Matt Taylor, the scientist in charge, was very much in demand.

'What was it like to find yourself thrust into the media spotlight like that?' Jim asked.

'It's something I'm not yet comfortable with.' he said. 'Let's put it that way.'

'Is it true that the European Space Agency advised you to cover up your tattoos when you were doing TV interviews?' Jim asked.

'It was more when we were doing media training,' Matt said, diplomatically. The ESA PR team had suggested that perhaps he should wear a suit. Matt, who had worn plenty of suits in the past, argued that they made him feel too hot. A brief discussion about his extensive body art followed. Perhaps he would like to wear a long-sleeved shirt? But Matt was keen to be himself. 'If I was going to be doing the talking, I wanted to be who I am,' he said. 'And now there's no point in me hiding my tattoos because everyone knows who I am. Maybe it's a good way to disguise myself. Maybe next year, after *Rosetta*, I can shave all my hair off, shave my beard and start to wear long sleeves, and nobody will recognise me. I'll go for your hairstyle, Jim!'

'If I was going to be doing the talking, I wanted to be who I am,' he said. 'And now there's no point in me hiding my tattoos because everyone knows who I am'

What ESA didn't spot (or, if they did, failed to do anything about) was Matt's choice of shirt at the press conference soon after *Philae* landed – a brightly coloured garment with a repeating motif of scantily clad women.

'Did anyone in the ESA PR team, or anyone else for that matter, suggest that maybe you should change your shirt?' Jim asked.

'I'm a little bit uncomfortable talking about this,' Matt said, guarded. 'I made my statement that week ...

It was me, basically. I made a mistake and I subsequently apologised. Do I regret wearing that shirt? I guess so.' He sounded unconvinced. 'Because of what happened, because of the noise and the negativity around the shirt, it's natural for me to say I shouldn't have worn it,' he said. 'There are other things I'd like to add, but not now. Maybe one day.'

I later learnt that Matt had chosen his shirt quite carefully that day. It was a present from his wife.

Matt's parents pushed him to work hard at school. They wanted him to go to university, so that he could have access to careers that had not been open to them. 'Certainly, I was led to believe that doing well at school would provide me with different opportunities, better opportunities later in life,' he said. 'And ultimately, I'm the proof of the pudding that it does pay off if you knuckle down.' It's something he constantly tells his own teenage children.

When he was a teenager himself, he used to go round to his uncle's house for tea. 'We'd go and have a few beers with him and chat about life in general.' His uncle believed 'physics was the way forward'. It was the science of everyday life, which sounded interesting to Matt. It was also the subject that his father and uncle thought would deliver the broadest base of options later in life. So he went to Liverpool University to read physics and his father, a bricklayer, got him jobs on building sites in the holidays so that he could pay his way. 'By the end of the summer I was counting in twelves, because each hob carried twelve bricks'[1] and ready 'to get back to the

1 But judging by a wall he built recently in his garden, his dad didn't show him how to lay bricks'!

books'. A few months of manual labour, carrying bricks up and down ladders and sweeping up, 'was a good way of remembering why I was at university'.

After graduating he joined the Space and Atmospheric Physics Group at Imperial College, London to do a PhD, having chosen them because he 'liked the vibe'. 'Or maybe they were the only ones who would have me,' Matt said, laughing. 'I don't recall.'

Jim suggested that plasma physics was perhaps less glamorous than other areas of astronomy – a suggestion Matt refuted. 'Plasma physics permeates everything,' he said. 'The easiest way to describe it, and this is something I constantly do with my mother, is to talk about it as the science behind the aurora.'

'Plasma physics permeates everything'

He got a job with ESA in 2005 to work on the *Cluster* mission, 'a pure plasma physics mission to investigate the space weather around planet Earth, looking at how the magnetic fields, the electric fields, the particles of plasma interact with one another.' These charged particles of plasma create 'a big tangled, tangled mess' that Matt and other plasma physicists try to unravel and understand.

In 2013, he was asked to look after all the scientific investigations on the *Rosetta* mission to comet 67P. Joining the mission seven years after it had launched, and 'after there had been some upheavals in the team', was a tough call. Perhaps feeling a little insecure about stepping in 'right at the end, effectively', Matt wanted to convince his new team that he was now committed to *Rosetta*, despite having lived with *Cluster* for a decade. 'I thought, maybe stupidly, that one way of proclaiming my dedication to the project was to have this tattoo done on my thigh.' He already had a tattoo of *Cluster* and thought Rosetta deserved one too.

'A bit like when you have a tattoo of your ex-girlfriend and your new girlfriend has to have one too?' Jim suggested.

'Yes,' Matt replied. 'I think that's what my wife thinks of the space missions I work on too!' Space missions, like partners, can be demanding.

'You once described *Rosetta* as the sexiest thing alive, second only to your wife,' Jim reminded Matt. 'Is it still as sexy as ever for you?'

'Um . . . it's funny. I find it more attractive when I'm doing outreach work and trying to explain it to other people.' The day-to-day reality is more mundane.

According to fellow space scientist Monica Grady, who developed one of the instruments on *Philae*, being the Project Scientist on a mission is 'a bit like being the *pater familias* of a great big family of dysfunctional teenagers'. Matt agreed but tactfully added that he didn't think some of the principal investigators on the mission would be very happy about being described in this way!

'So is there stomping off and slamming of doors?' Jim asked.

'Of course,' Matt said, smiling. Balancing multiple and often competing demands on *Rosetta* to fly a bit closer, or tilt as much as possible in a particular direction, was a challenge. The precise route taken by the spacecraft during each flyby of the comet had been hotly debated, with each of the principal investigators wanting to ensure the best possible operating conditions for their instruments. Often it is against one PI's interests 'to even support something from somebody else'. If they miss out on an opportunity to get data from a comet that is 4 billion miles away from planet Earth, 'they are never going to get that opportunity back'.

And then there were the engineers. PIs are all scientists by training. They want to squeeze as much information as possible out of the mission, 'that's why they are there'. They want to maximise their data. 'They want to do things that are perhaps too extravagant,' Matt said kindly. The operational engineers who drive the spacecraft have different concerns: they need to 'push a very large spacecraft with 64 square metres of solar panels towards a body that is constantly trying to push it away. It's like going down to the beach on a winter's day with your anorak open and trying to walk into the wind.'

The engineers need to make sure the spacecraft survives intact. At the same time, they sometimes need reminding that without the scientific imperative, 'they wouldn't have such a fantastic machine to be flying around.'

The operational engineers and the scientists have very different agendas, and Matt spent a lot of time and effort trying to prevent the two groups from going head to head. 'I'm kind of in between all that, trying to filter out some of the noise.' Without Matt in the middle, to absorb the occasional angry feelings on both sides, 'nothing would get done'. Someone somewhere would whip out a yellowing document from their back pocket and start to whinge: 'You said this five years ago . . .' And so it would go on.

Jim sympathised with the engineers. 'They don't want their fantastic machine to break.'

'In a way, they don't want any instruments on board,' Matt said, 'because they know the scientist will get in the way of them flying the spacecraft perfectly. Their day-to-day engineering performance is disturbed. That's it. It's a tension.'

There was a close call when *Rosetta* was about 6km from the surface of the comet. They very nearly had to put all the instruments into safe mode, 'to get out of Dodge, as it were'. The instruments were at risk of being destroyed and had that happened, there would have been no more data from the mission. 'We were lucky we didn't go into safe mode. We didn't lose any signal as we were going through this close approach, but it could have happened. Anything in space can go wrong. That's the thing. If you do something a hundred times, the hundred-and-first time you do it, it might go wrong. All the time you are worried, but you can't let that weigh on your mind, because you've got other things to think about and plan. You have to plan positively.'

'What's been your darkest moment so far on the *Rosetta* mission? Jim asked.

'I think generally around the landing,' Matt said. 'I have very clouded memory of the whole week when *Philae* landed.' But he does remember the relief he felt 'when everything returned to normal'. There were no more media interviews and he was sitting with his laptop on the coffee table at home, quietly sharing the latest findings from *Rosetta* with all the scientists involved, updating everyone on what had been discovered.

'The worst part of the week for me was the day before the landing. We had some problems with the lander.' There was an emergency meeting and Matt signed off a set of criteria that would need to be met before *Philae* would be ready to leave the mothership and begin her descent to comet 67P. It was the first time he had seriously considered the possibility that things might not be able to proceed as planned. 'Up to that point I hadn't conceived of a no-go situation,' he said. 'And there it was, in the cold light of day, or rather the cold light of 11.30 at night. What would happen if we couldn't get the "go" criteria ticked off?'

He went to the bar in a hotel near mission control in Darmstadt, Germany with his colleague Fred, but it was full of persistent people hell-bent on getting answers to their questions, so they moved to the smoking area which, strangely, was journalist-free. 'There didn't seem to be many smoking journalists,' Matt said. The ESA communications team wanted to run through some possible scenarios and 'it was getting quite frightening', but after talking things through, he told everyone that everything would be clear at 3.30 a.m. and went to bed. 'There was no point in worrying' because there was absolutely nothing that could be done. Easier said than done, perhaps.

At 3.30 a.m. precisely, he was woken by 'the quietest beep'. A text from Fred. 'We're go,' it said. Later that morning *Philae* was released. She bounced and bounced again before settling on the surface – or, as Matt described it, *Philae* 'did this little journey across the comet'.

Three days after landing, *Philae* shut down. Perhaps she was in a deep crevice which blocked out the sun's rays or perhaps she had landed on her side. Either way, she had no access to solar power and her batteries had run down. *Rosetta*, however, continued to orbit the comet and send back pictures.

'When those first images of comet 67P came back, were you surprised to discover that it was shaped more like a duck than a potato?' Jim asked.

'As a plasma physicist, I don't have a great appreciation of pictures,' Matt said. He was pleased, nonetheless, to learn that comet 67P had 'all the best features of all the best comets'. Strange shapes, boulder-strewn terrain, collapsed cliffs and deep pits had all been observed from a distance on other comets. Comet 67P had all of these exciting features, not just one or two, offering us an opportunity to study them in more detail than ever before. And, yes, the duck-like shape was a big surprise. 'It was striking. It was a really nice present.'

'What sort of a place is it?'

'I always get mixed up. Is it a dirty, dusty snowball or a snowy dustball?' Matt said. Probably the latter, because recent measurements seem to suggest that there's a lot of dust coming off the surface. 'It appears to have these crusty organic layers. It's very dark. The albedo [the proportion of incident light that is reflected by a surface] is about 4 or 5 per cent.' (Planet Earth has an albedo of about 30 per cent.) And *Philae* has detected what seems to be a very hard subsurface. 'There are indications that there's ice under there, and this is what we're kind of waiting for. To see how this evolves.'

'Why does it matter?' Jim asked, having apologised in advance for asking such a heretical question. 'Why is it important to understand the properties of a lump of rock and ice billions of miles away from planet Earth?'

'Well, comet 67P has been on the outskirts of our solar system for a very long time.' The material it contains was put into a deep freeze right at the beginning of our solar system. 'That's why it's interesting,' Matt said. And complex organic molecules have been found.

Jim asked Matt if he was shocked to find that such complex organic molecules existed 4.5 billion years ago.

'It was pleasing,' Matt said, 'because these complex molecules are the building blocks of life.' And if there were complex organic molecules present in the earliest days of our solar system, when the planets were being created, it

'Why did life evolve only on Earth? What was it like back then and why are we here? We're touching on those big questions'

begs the question: Why did life evolve only on Earth? 'What was it like back then and why are we here? We're touching on those big questions,' Matt said.

'Do you think this makes it more or less likely that there is life elsewhere, whether on one of the moons of the outer planets or in other solar systems?' Jim asked.

'It's strange that we haven't found anything yet,' Matt said. 'We've seen that the building blocks were there.'

'There has to be something somewhere else,' Jim suggested.

'I hope so,' Matt said.

When Jim interviewed Matt, *Rosetta* was still sending back data from comet 67P and *Philae* was still out of action. The team, meanwhile, was keeping the faith, 'just waiting for it to come out of hibernation'. Matt had calculated that the best illumination conditions would be in May 2015 and was hopeful that, exposed to sufficient sunlight, *Philae* would wake up. She would, he said, probably behave rather like his teenage daughter when she tells him repeatedly that she will be up 'in a minute', before finally getting out of bed. 'Hopefully, by the third time you shout up the stairs your teenage son or daughter gets out of bed.' Similarly, he expected that if the sun did shine on *Philae*, she might stir several times, sending back the odd signal sporadically, before finally springing into action.

'What are you most worried about at the moment?' Jim asked. 'What keeps you awake at night?'

'I don't have trouble sleeping,' Matt said. 'I just don't sleep much.' Having analysed the results generated from the first round of data, he was keen to 'react to lessons learned', but the mission was already operating at maximum capacity and so introducing new ideas felt risky. 'Everyone is working 24/7 and so making a change could push things over the edge.' It's a fine balancing act between making the most of being there and risking bringing the whole mission to a premature end.

Finally, Jim asked Matt if he had ever been tempted to go into space himself, rather than relying on data to find out what it's like.

'The last astronaut selection programme, the one Tim Peake got selected on – I was contemplating applying,' Matt said. He was just young enough. But the price of the compulsory pilot's medical certificate put him off. 'I thought I could get a PlayStation instead,' Matt said, laughing. 'Or a bike.'

Courtesy of Nick Fraser, National Museums of Scotland

NICK FRASER

'I was really out of my depth going back another hundred million years'

Grew up in: Newhaven, Sussex
Home life: married to Christine with two daughters, Hannah and Amy
Occupation: palaeontologist
Job title: Keeper of Natural Sciences, National Museums of Scotland
Inspiration: fossilised bones emerging from a block of sandy limestone
Passion: ancient reptiles
Mission: to understand our evolutionary past
Best moment: spotting the ribs of a Triassic reptile, buried in a rock and glinting in the sun
Worst moment: accidentally putting a pick through the skull of a fossilised whale
Advice to young scientists: observe the world around you
Date of broadcast: 11 April 2017

Nick Fraser is a time traveller. He regularly travels back in time (at least in his mind) to the Triassic, a crazily inventive period in our evolutionary history between 250 and 201 million years ago when ancient reptiles, some the size of small cars, dominated the land, sea and sky. For 18 years, he worked at the Virginia Museum of Natural History, USA, and spent a lot of time hunting for fossils in the nearby Solite quarry, a treasure trove of Triassic creatures, many of which look utterly bizarre to modern eyes. In 2002, he unearthed a lizard-like reptile with wings and an extraordinarily long neck, the first example from a previously unknown, and now extinct, group of animals, Protosauria. Since then, he has discovered more reptiles in this group in southern China. He moved back to Scotland in 2007 and was involved in another exciting discovery much closer to home.

As a child Nick Fraser played with woodlice, carefully transporting them around the house, while his mother looked on in horror. A few years later, he was reading *On the Origin of Species* by Charles Darwin. 'I felt like I had to,' he said.

His parents had bought him *The Larousse Illustrated Encyclopedia of Animals* (a more age-appropriate book, perhaps) and he used to sit on the floor, reading. 'I remember doing incredibly detailed drawings of all the animals,' he said. 'And it got my imagination going about what's in the world.'

Impressed by the diversity of life, he wanted to know where all the different animals had come from and hoped Darwin would provide some answers. 'Did I understand it?' Nick said. 'Probably not that well, but it was well thumbed. I could see there was a passion from the author. That's what sparked me.'

'You've spent most of your career studying creatures that died out hundreds of millions of years ago, but at the beginning of your Life Scientific, as you've mentioned, you were more interested in living things,' Jim said. 'So what was it that changed?'

'I went to study zoology at Aberdeen University and I got really excited by parasites,' Nick said, 'especially malaria.' After graduating, he wanted to study malaria at Glasgow University, but the reserach funding he needed was on a two-year cycle, and he had to find a job to fill in. When he started working at the Geology Museum in Aberdeen, he knew 'very little' about fossils, but witnessing tiny, ancient bones emerging from lumps of sandy limestone (that had been brought back by students who had been on a field trip to the Mendips) turned his attention from the parasitic living to the long dead. He was soon addicted to the process.

'You take a bowl of acetic acid,' Nick said. 'Vinegar, very strong vinegar, and you dissolve the rocks for a bit.' The limestone would fizz like Alka-Seltzer in these acid baths, as the calcium carbonate reacted. Fossilised bones, made from a different calcium salt, were more resilient. 'Then you wash them out and drain off the sediment. And just keep repeating this day in, day out, collecting all the bones. And then you end up with this pile of bones,' mixed with sand and other acid-insoluble sediment. Confronted by heaps of tiny, disarticulated bones, he set to work, sifting and sorting, hoping to recreate plausible creatures, with only the vaguest sense of what they might look like,[1] and no idea how many animals were even present in his hotch-potch collection of bones.

'How easy was it to fit the bones together?' Jim asked. 'I'm rubbish at jigsaws, so I'm always forcing pieces in. "Try it here for now." Do you do that sort of thing with these bones?'

'These bones are very, very fragile, so there was no forcing these pieces,' Nick said. He would nudge them along with a wet paintbrush, experimenting with different arrangements of bones to see which made the most plausible skeleton, in the days before there were computer programs to help with this kind of thing. 'I could see

1 Initially Nick assumed that everyone else would know more about these bones, and which creatures they belonged to, than he did, and he sought advice before trying to put them together. The he discovered that 'nobody really knew what they were or where they came from', which only made him even more interested.

different sorts of teeth, different sorts of jaws.' It was as if someone had taken the pieces of maybe 20 different jigsaw puzzles and thrown them all together, 'and thrown a few pieces away as well', and then said, 'Sort them out!'

'I could see different sorts of teeth, different sorts of jaws'

'So, twenty different jigsaws, all in the same box, and these are 3D jigsaws, I guess,' Jim said. 'There's no picture to guide you and you don't even know how many different pictures there are.'

'When I started I didn't have a particular picture in mind,' Nick said. He started to look at the palaeontological texts and found Alfred Romer's *Osteology of Reptiles*[2] very helpful. Teeth, he learned, were a good place to start, as they tend to be very species-specific.

'Presumably you get left with lots of odd bits of bones that don't go anywhere?' Jim said. The individual lumps of rock that Nick was studying don't respect the integrity of the animals buried within them. It's just the way the limestone crumbles.

'Yes. And those are probably the most interesting ones!' said Nick, happily exasperated. 'Those are the ones that, in twenty or thirty years' time, someone will come along and say: "What an idiot that guy was! He should have seen that here was the first . . . whatever."'

As he worked away in his vinegary teaching lab in Aberdeen, Nick was piecing together tiny reptiles that lived 220 million years ago. They dated back to a time known as the Triassic, when the rocks now buried in the Mendip hills in south-west England were the bottom of a

2 A classic textbook. More on Alfred Romer and Romer's Gap later.

tropical lagoon. A vision of warm, shallow waters teeming with Triassic life kept him going when the fossil jigsaws weren't going so well and propelled him to do a parasite-free, fossil-rich PhD on these tiny lizard-like creatures from the Triassic. Several post-doc research jobs on this under-studied epoch followed.

Then in 1990 he became the director of the Virginia Museum of Natural History, moving his wife and young daughter across the Atlantic, following the Triassic rocks westward. His friend and colleague Paul Oulson had described a nearby quarry as 'stupendous' and Nick was keen to do some digging. Several exciting Triassic finds had already been made from this previously little-explored bed of rocks and Paul's wide-eyed enthusiasm had been infectious. But Nick's first visit to this much-hyped quarry was a disappointment. Paul kindly showed him all the best places to look, and Nick found absolutely nothing.

'I was getting dirtier and dirtier, hotter and hotter, and couldn't see a single thing!' After hours and hours of picking up hot, dusty rocks and splitting shale, his disappointment grew. 'I did begin to think, "Is this it? Oh my goodness, Paul, what have you led me into?"' He had uprooted his family, changed jobs and continents and was experiencing a degree of culture shock, living in the American South, surrounded by Baptists. The neighbours made the family feel very welcome, but many didn't agree with Darwin about the origin of species.

Several visits to the legendary Solite quarry later, Nick started to 'see the richness' buried within boulders that were otherwise nondescript. And the weird and wonderful world of the Triassic opened up before him.

'It was a crazily inventive period in our geological history,' he said. After the Permian mass extinction wiped out about 90 per cent of all species, all manner of strange animals evolved that look utterly extraordinary to modern eyes. Reptiles ruled the land, sea and sky, 'a weird, eclectic group of animals with Heath Robinson-type contraptions.' Many were the size of small cars. And amid all these bizarre and crazy-looking ancient reptiles – better suited, perhaps, to the pages of a Dr Seuss book than an encyclopaedia published by Larousse – some more familiar species emerged.

There were mammals, including tiny, shrew-like creatures, as well as the first lizard-like animals, the first turtles and the first crocodiles, even if the last of these did look very different from their modern counterparts. 'Crocodiles weren't the sluggish, poolside couch potatoes that you see today. The first crocodiles were little, active, dog-sized animals.'

The first water bugs and ground beetles and approximately 30,000 other Triassic species, most of them insects, are still thriving today, 200 million years on. 'So here is the beginning of the modern world,' Nick said. And it was these 'modern' insect species that interested Nick most. Using a CT scanner,[3] he was able to discover thousands of tiny Triassic creatures, just-like modern day thrips (those tiny pests that eat garden roses and ruin orchards), squashed flat 'with all the little hair-like projections on the antennae still preserved.'

'The first crocodiles were little, active, dog-sized animals'

3 CT scans allow palaeontologists to see through rock and find fossils *in situ*. No digging or acid baths required. The scan removes the surrounding rock from view, just as it renders our flesh invisible during medical checks.

Once he had got his eye in, it was clear to him that 'The Solite quarry is, without question, the most important Triassic site anywhere in the world.'

'Did you find yourself travelling back in time in your mind and imagining the world as it was over 200 million years ago?' Jim asked.

'In the summer, the sun's beating overhead. The humidity is matching the levels of the temperature in Fahrenheit. You soon start going crazy. You get maybe a little delusional and you do start imagining things.' But, Nick said, a vivid imagination is to be encouraged, not dismissed as unscientific. It helps to be able to picture in your mind's eye how buried creatures might have lived. Splitting rock and seeing a fossilised creature trapped inside 'makes you think, what was this?' And when it's something that he has never seen before, then 'you do want to picture what it was like. So, yes, it's exciting. Even if you start losing your senses a little bit because you've spent a bit too much time in the sun.'

In the summer of 2002, Nick was in the quarry, as usual. A local news team was making a short film and he was showing the TV crew what he would be doing. He spilt a rock with a hammer, glanced at the two parts and was about to embark on the usual 'nothing-there-throw-it-over-your-shoulder' routine, when something made him pause. The sun caught the newly split face and Nick thought: 'Oh, that's a fish tail. OK, I'll keep that.'

Nick thought it was nice to have, but not extraordinary. Later that day he changed his mind. 'I took a closer look and thought, "That fish tail's got a neck on it and a skull! Mmm, that's not a fish." And there it was, this gliding reptile . . . Even then, we couldn't really be sure what it looked like,' Nick said. 'To this day we haven't got it out of the rock.' Sometimes it's just not possible to extract fossils, or the risk of trying is too great, so this fossil has been described entirely on the basis of CT scans. CT-assisted vision, however, can be better than the naked eye.

'It's not just a squashed fossil on the surface – you can turn it the other way round. You can see the underside, the side and all the details,' Nick said enthusiastically. 'You can start to get into the brain case!' And all the while fragile fossils are protected, safely encased in rock.

In this way Nick revealed a lizard-like creature, about a foot long, with wings and an extraordinarily long neck (not an obvious feature for a reptile that glides, and therefore surprising). Even by Triassic standards, this specimen was odd.

Nothing quite like it had been described before. Which prompted the question: if this previously unknown group of animals has just come to light, what other weird Triassic reptiles might still be waiting to be found?

At the same time, in southern China, yet more wacky creatures from the Triassic were being unearthed. Bulldozers were churning up farmland that had been undisturbed for centuries, replacing crops with breeze blocks, and Chinese palaeontologists had moved in to see what they could find. Predictably, perhaps, Nick started escaping to China to investigate whenever he could, and met Li Chun, a teacher who hunted for fossils in his spare time. He was 'an amazing gentleman', Nick said, 'who would sort of lean over conspiratorially and say, "I've got something special to show you,"' and who would share the most extraordinary fossil finds. 'And the best thing is, I've seen these fossils being collected,' Nick said, excited. 'I've seen them with sediment over them and watched specimens emerge. These are not forgeries! They are the real thing. And it's just stunning!'

Li Chun showed Nick the remains of animals with vacuum-cleaner-like nozzles that, despite his extensive knowledge of crazy Triassic creatures, made him think, 'What on earth is this?'

'These reptiles are blowing apart our ideas of evolution,' he said. 'The Triassic was just such a wild and wonderful time.' And it wasn't easy working out what the creatures looked like, 'because these things have been flattened, severely flattened'. To get some idea of the challenges involved, try imagining what a wardrobe might look like by studying a box of flat-packed pieces, many of which are battered, chipped and crushed.

'These reptiles are blowing apart our ideas of evolution'

The reptiles found in the Solite quarry in the USA were land animals, but Li Chun's specimen clearly lived in the sea. It had a strangely elongated snout, and it reminded Nick of a marine reptile that had been discovered in China a few years before (in 2014), named *Atopodentatus*. (*Atopos* is the ancient Greek word for unplaceable, eccentric or disturbing.) 'It was just such a weird animal,' Nick said.

The scientists who first discovered this strange species of marine reptile found a specimen with a poorly preserved skull. They had assumed (reasonably enough) that this reptile passively took in nutrients from the sea as it swam along, and they had assembled what bones they had to construct a conventional-looking mouth and snout. But Nick wasn't convinced. The chisel-like teeth in the specimen Li Chun had found were made for biting and scraping, not passive filter feeding, he thought.

Proving the existence of the Higgs Boson cost millions, with thousands of scientists around the world each devoting decades of their lives to proving that it did indeed exist. When Nick wanted to test his ideas, he drove to the local supermarket in Beijing and bought a pack of plasticine and some toothpicks.

The question that Nick wanted to answer was this: what did these reptiles eat and how? 'We want to understand exactly how they lived, and that's the exciting bit,' he said. Impatient to find out, he stood in the supermarket car park with Li Chun, making plasticine models of possible jaws, breaking toothpicks in half to represent teeth, modelling and remodelling possible mouth shapes for *Atopodentatus* until they had created a model jaw that was capable of eating.

'We want to understand exactly how they lived, and that's the exciting bit'

With the snout arranged as it had been described in 2014, a large gap appeared in the middle of the snout through which water would simply slip away and, with it, all hope of food. Nick's conclusion: a filter feeder with a mouth like this would have been very hungry indeed.

Instead, Nick and Li Chun put the elongated end of this reptile's mouth at right angles to the rest of its snout. It certainly looked quirky (like a bulldog clip on the end of a narrow pole). But it was the only arrangement of the flattened, fossilised remains that seemed to be fit for purpose when modelled in three dimensions. If this reptile had a hammerhead, it would be able to eat by actively biting off sea plants rather than passively ingesting plankton and small fish. Nick could imagine its protruding snout clamping shut around some swirling algae and the chiselled teeth scraping a nice mouthful of seaweed off the rocks. Then, gulping, it might squeeze the excess water out. The net result: a reptile with an odd-shaped head gets a nice square meal.

Marine reptiles that were vegetarians were unheard of in the Triassic and are still extremely rare. But the plasticine-and-toothpick test convinced Nick and Li Chun that this eccentric-looking reptile, with its fearsome hammerhead, was only capable of eating plants. They published a paper describing this 'highly specialised feeding adaptation' and *National Geographic News* reported: 'Ancient Reptile Ate like an Underwater Lawn Mower'. The bizarre snout design described by Nick and Li Chun is now widely accepted and *Atopodentatus* has been identified as an early and rare example of a marine herbivore.

'You are now, and have been for some time, a world expert on the Triassic. So what was it that persuaded you, about ten years ago, to travel back in time another hundred million years to a time long before the reptiles that you know so well?' Jim asked.

'I was really out of my depth going back another hundred million years,' Nick said. After 18 years in the USA (which included several trips to China) Nick moved back to Britain to become Head of Natural History at the National Museums of Scotland in 2007 and, sitting at his desk in Edinburgh one day, he received a phone call from an old friend.

'Stan Wood phoned me and said, "I may have something you could be interested in," and knowing Stan from 30 years ago, I said, "I'll pay attention to this." So we met and he showed me something.'

Stan had spent a quarter of a century searching for fossils to fill Romer's Gap, an infuriating 15-million-year gap in the fossil record (between 360 and 345 million years ago) which pops up just when some of the most exciting evolutionary action must have taken place. At the beginning of Romer's Gap four-limbed vertebrates, known as tetrapods, were all swimmers. They had gills, fishlike tail fins, and limbs with more than five digits. By the end of the period, the tetrapods had mastered the art of living on land. The fossil remains of these later creatures reveal lungs and robust limbs, and there is no evidence of fishy tails. So at some point in Romer's Gap we can be fairly sure that limb-like fins became fin-like limbs and gills became lungs. And as a consequence of these changes, tetrapods evolved to survive on land for more than an occasional slither around.[4]

Stan phoned Nick to tell him that he had found the fossilised remains of several tetrapods with lungs, near the Whiteadder River in East Lothian. It was a discovery that was too exciting for Nick to ignore, even if moving out of his geological comfort zone made him feel a little insecure. Nick alerted Tw:eed (Tetrapod World: early evolution and

4 Alfred Romer, a great specialist in vertebrate evolution, spent most of the twentieth century searching all over the world, and in vain, for fossils to fill the gap that bears his name. Some thought, rather arrogantly perhaps, that if they couldn't find any fossils of this age, then none existed, and so they attributed this surprising lack of life to lower-than-average oxygen levels in the Earth's atmosphere during Romer's Gap. For Stan there was only one conclusion: *Must try harder*.

diversification), a large-scale scientific collaboration set up by Jenny Clack to study rocks in the Scottish Borders, the North of England and Northern Ireland that date back to Romer's Gap. And, on his advice, four large blocks were carved from the rocks near the Whiteadder River where Stan had made his find.

On the surface of one of the rocks Nick spotted what he thought was a lungfish. He knew from experience that where there are lungfish, there are often early amphibians, and he was keen to learn more about them. But when he did a CT scan of the rock, it revealed something unexpected. Hidden among what he had thought were lungfish bones was something altogether more exciting: the skull of a small amphibian.

'Oh my goodness, forget the lungfish!' he thought. 'Inside this piece of rock we could see, in the CT scan, this little amphibian.' With a head that was just 4cm long, it was christened Tiny. (Most of the ancient amphibians that were known at that time were the size of dogs.) But the scientific name is perhaps more poetic: *Aytonerpeton microps*, 'the creeping one from Ayton with the small face'.

'Does this discovery mean that there could be thousands of fossils, well hidden, that even five or ten years ago we would not have been able to see?' Jim asked.

'Yes. It's a new tool,' Nick said, 'and who knows what tools we will have in the future?' Maybe soon, fossil hunters will be able to CT scan a whole region of rock and there would be no need for people like Stan to put their waders on and excavate.

'Oh, that's really cool!' Jim said. 'And, of course, this is such an important period in our evolutionary history.'

'These are the animals that first came out of the water, the first vertebrates to emerge with four limbs,' said Nick.

The moment Tiny or another, similar specimen appeared on land is a Neil Armstrong moment: 'One small step for tetrapods, one giant leap for vertebrate evolution.'

Tiny has lungs, a backbone, four limbs and, most probably, five fingers. And we all know how useful that particular combination of anatomical assets can be.

'It is a very, very important time period,' Nick said. 'I hesitate to say it's more important than the Triassic.' He laughed. 'But it is very significant and it is right here on our doorstep. Without Tiny there would be no birds, no dinosaurs, no crocodiles, no mammals, no lizards and obviously, we wouldn't be around.'

Several other significant tetrapods have been found in the rocks near Tiny and Tweed excavations continue. It is these finds in the Scottish Borders that are helping to close Romer's Gap. Meanwhile, the search has widened to include other rocks of a similar age that can be found in China, Antarctica and Greenland, East Greenland in particular.

'We want to see what might come out of rocks of the same age from those areas,' Nick said. 'Were the borders of Scotland the cradle of evolution for life on land, or was it somewhere else? This is the first time we have been lucky enough to find something. Was that because these rocks are where it all started? Or was it somewhere else? That's the question. That's the quest.'

Imperial College London

MICHELE DOUGHERTY

'I can look at this page of data and see where the spacecraft is'

Born in: Kwazulu-Natal, South Africa
Occupation: space physicist
Job title: Professor of Space Physics and Principal Investigator for the magnetometer instrument on the *Cassini* spacecraft
Inspiration: 'My dad, who instilled in me the idea that I could do anything I wanted to do'
Passion: interpreting data on magnetic fields in space
Mission: to explore Enceladus, a small icy moon of Saturn
Best moment: 'Confirming that there were plumes of water vapour coming off Enceladus'
Advice to young scientists: 'Have confidence in yourself. Don't be put off by minor hurdles'
Date of broadcast: 10 January 2017

Michele Dougherty grew up in rural South Africa, surrounded by fabulous night skies. She helped her dad build a telescope in the garden, but it was studying mathematics later in life, not seeing the rings of Saturn as a child, that pulled her into space. For 20 years she lived, breathed and worried about the *Cassini* mission to Saturn, the most ambitious space mission of the twentieth century. The mathematics she created to understand data on the magnetic fields in space won her the prestigious Royal Society Hughes Medal in 2008, joining former winners such as J. J. Thomson, Alexander Graham Bell, Niels Bohr and Peter Higgs and just one other woman, Hertha Ayrton in 1906. During a distant flyby of Enceladus, one of Saturn's smaller moons, she spotted an anomaly in the data and, in a bold move, she persuaded mission control to divert the spacecraft to take a closer look.

When Michele's boss, David Southwood, asked her to take his place as a Principal Investigator on the *Cassini* mission to Saturn, she was excited and 'absolutely terrified'. There were many highly experienced scientists in her team. 'All of them were men.' She was a mathematician in her mid-thirties with no previous management experience.

'Why do you think he chose you?' Jim asked.

'I think he saw something in me that I hadn't seen myself,' she said. 'I think I'm quite good at bringing people together as a team and helping them see an end goal.'

When Jim asked Michele if she was practically minded, she laughed. 'I never felt comfortable in the lab.' One day at university she turned the computer on and 'smoke came billowing out of the back . . . And I thought, "I can't even touch computers without them blowing up." So it got to that stage when I was told I wasn't allowed to handle the experiments themselves. My [lab] partner would do that and I would just take the notes.'

'And here you are, responsible for one of the most expensive instruments,' Jim said, smiling.

'Absolutely!' Michele said, laughing. The main instrument is safe, a billion kilometres away on board *Cassini*, but there's a flightspare instrument in the lab back at Imperial College London which is used to test out commands. 'And I've been told by the instrument manager that it's probably best if I leave him to do that job,' Michele said. 'Because no one trusts me near the flightspare, I'm afraid.'

Michele grew up in South Africa during the apartheid era. She studied mathematics and biology and 'a few other subjects' at school, and went on to study at the University of Natal (now the University of Kwazulu-Natal). 'My dad worked at a university and I was able to get free education as a result.' She chose to study science 'just to see if she could', since neither physics nor chemistry had been options at her girls-only school. It wasn't easy. 'The first year was really hard,' Michele said. 'I hadn't done physics. I didn't understand any of the concepts. And so I'd go home every evening and my dad would go through the lectures with me.' Throughout her undergraduate degree and subsequent masters she felt out of her depth academically, despite performing well. 'I never really felt as though I caught up until I was doing my PhD.' Writing a thesis on applied mathematics, at last she felt more in control.

After a few years at the Institute for Theoretical Physics in Heidelberg, she got a post-doc job at Imperial College and moved to London. 'And there, for the first time, I started looking at data,' she said.

Colleagues at Imperial asked Michele if she wanted to spend a little time putting together a magnetic-field model to help interpret data from the *Ulysses* mission to the sun. In order to gain the necessary momentum to go into orbit around the sun, the spacecraft needed to follow a desperately indirect route, flying from the Earth to the sun via Jupiter 'to give it the kick it needed'. And when mission control at NASA's Jet Propulsion Lab (JPL) were monitoring the Jupiter flyby, they invited Michele to join their data-analysing team in Pasadena, having heard about her mathematical work, modelling magnetic fields in space.

Walking into a room full of scientists who looked very self-assured, her own confidence drained away. 'What am I going to do?' she thought. 'They all know a lot more than I do.' So she sat there quietly, hoping no one would ask her any questions. 'Slowly but surely I realised that I knew a little bit more than I had thought and became really enthused about working with data, understanding what the data could say.'

Working out how to interpret the magnetic-field data that was being sent back by *Ulysses* was a steep learning curve, but when Michele cracked it, the insight into events in space that the data gave her was more exciting to her than the view of Jupiter through her father's telescope, impressive as that had been.

'Was that the moment you transformed from a mathematician, an applied mathematician, into a space scientist?' Jim asked.

'I think so, yes,' Michele replied. 'I realised you could use these wiggles on a page to understand something that was happening five astronomical units away . . . This was a sea change.'

'I realised you could use these wiggles on a page to understand something that was happening five astronomical units away'

After working with the data for a while, Michele was able to look at a page of numbers and tell where in space the spacecraft was. And it became clear that she could use the skills she'd picked up in her PhD to help her analyse the data. The mathematical models Michele had created in London told her what the magnetic field around Jupiter should look like, undisturbed. The incoming data on Jupiter's magnetic field measured what conditions were like with a spaceship present. By comparing the measurements from *Ulysses* with the measurements predicted by her model, she found a way to calculate where *Ulysses* had to have been in order to create the precise disturbance that had been observed. The mathematical techniques Michele invented to track *Ulysses* were both ingenious and enormously useful. Armed with insights like these, she no longer feared not having anything to say.

As a result of this work, David Southwood invited Michele to join the team on the upcoming *Cassini* mission to Saturn. Soon after, he asked her to take over from him as Principal Investigator on the magnetometer team. The magnetometer is akin to a highly sensitive compass. It measures the magnetic fields in space and is one of the most important instruments on board *Cassini*. So it was a big promotion. The first couple of years were 'a little bit difficult'. Being responsible for a team of 40 highly experienced scientists, who were single-mindedly pursuing different research goals, was a challenge.

'Was it tough?' Jim asked.

'The trick is to keep everyone equally unhappy,' Michele said, and laughed. 'Joking aside, you've got to make people realise that they are part of a team, and that their work is not necessarily the most important.' By accident, she discovered her most effective management technique. 'One of the co-investigators pushed back on me really hard,' she said, 'and I just lost my temper. I didn't mean to. I don't like losing my temper. But it worked like a charm.'

The fun, and sleepless nights, began when *Cassini* went into orbit around Saturn in July 2004, having launched in October 1997. Seven months later it performed a flyby of Enceladus, a small and, at that time, apparently unimportant moon of Saturn.

'What were you expecting to find?' Jim asked.

'Nothing!' Michele replied. It was a distant flyby, and being 1,000km from the surface of a moon that itself was only 500km wide, the team wasn't expecting to see anything at all. 'And I have a confession to make,' Michele said, as if it were an act of gross negligence: 'We didn't look at the data for about 24 hours afterwards.' They were busy doing other things.

When they did look at the data generated by the magnetometer during this distant flyby, Michele spotted some unexpected 'bumps', extra signatures that were not easily explained. At first, she assumed she had been careless, blaming herself for this apparent error. Maybe they hadn't updated the altitude of the spacecraft during the flyby? *Cassini* had been tilting and pitching quite a lot to try and get the best possible pictures of Enceladus, movements that could easily have been overlooked.

The magnetometer team checked over all the data, looking for mistakes, and found none. And so the inexplicable 'bumps' remained, prompting the question: why were they there? The data suggested, when analysed with Michele's mathematical techniques, that something was protecting Enceladus from the magnetic field surrounding Saturn. 'It was as if Enceladus was much bigger than we thought,' she said. But 'we weren't quite sure what we were seeing, so we kept quiet.'

Another distant flyby of Enceladus was planned for a month later. The magnetometer team waited eagerly to see what the data might reveal this time. A moon that just a day ago had been of no real interest to anyone was suddenly intriguing, and they were the only ones who knew.

The magnetic-field activity that the data seemed to describe was not what you would expect from a small spherical moon, and the most sensible explanation led Michele to the most unlikely conclusion. 'It looked like an atmosphere,' she said – not the kind of breathable atmosphere that we are used to here on Earth, but a protective gas surrounding the surface nonetheless.

'And having an atmosphere around this small moon was an unusual thing?' Jim asked.

'Absolutely!' Michele replied. 'It's so small that you wouldn't expect the atmosphere to stay there long.'

Small planetary bodies lack the necessary gravitational pull to hold onto anything remotely atmosphere-like. An atmosphere around such a tiny moon just didn't make sense unless, Michele reasoned, something was replenishing it constantly. Something must be actively generating this unlikely atmosphere. 'That was the real interest,' she said.

The second flyby confirmed what they had seen before. It detected the same inexplicable 'bumps'. And a further observation added to their excitement. Another highly atypical pattern emerged from the data analysis: 'little waves' that looked as if they were created by ionised water-vapour particles. And this meant 'there was water, liquid water maybe?' This was a surprise. No one had expected to find water on Enceladus. No one had expected anything at all from this tiny and, apparently, insignificant moon. Now Michele and the magnetometer team were convinced that Enceladus was exciting and were desperate to explore further. But since not everyone had seen what they had seen, the challenge was persuading everyone else that their observations were significant and needed to be followed up.

Cassini was due to perform a third distant flyby of Enceladus in July 2008, but Michele wanted to go in close. She resolved to try and persuade mission control to divert *Cassini* from its well-established route. It was a bold move to upset plans that had taken years to put in place and it added several new elements of risk. She flew over to California for a big *Cassini* science meeting at the NASA JPL. Just before the meeting, she went to buy a coffee, which she 'probably didn't need', and started chatting to the man in front of her in the queue, who just happened to be the chief engineer in charge of *Cassini* at the time, Jerry Jones. When Jerry asked Michele why she was in Pasadena, she told him about her desire to fly *Cassini* as close as possible to Enceladus. Such a plan, Jerry knew, would use up much-needed fuel and risk turbulence, but, most unexpectedly, he responded by saying, 'That would be cool.'

The 'flight guys', as they are known, are professionally risk-averse. It's their job to keep the mission on course, but apparently Jerry 'had always wanted to take a spacecraft closer to a planetary body than anyone else' (even if admitting this to his boss would be to risk losing his job).

Michele went into the meeting buoyed up by this unexpected enthusiasm from the engineers and turbo-charged by a surge of optimism. The scientists, however, still needed to be persuaded. 'You show people data and their eyes glaze over,' Michele said. So she found another way to communicate and converted all the data into a simple schematic. If a picture is worth a thousand words, a visual

'You show people data and their eyes glaze over'

is probably priceless when it comes to representing what pages and pages of numbers actually mean.

Michele made her case. She showed everyone her schematic and said, 'I think we're seeing something really exciting. Could we go close?' As expected, there was 'quite a lot of push-back'. Several scientists mentioned, forcefully, that they had spent six and a half years planning 'every second of the voyage'. It made a mockery of the whole process if those carefully arrived-at plans could be changed, apparently on a whim.

Everyone wanted to skew the spacecraft in their favour, often quite literally. Sometimes the whole spaceship had to tilt in order to achieve a favourable angle for the camera or one of the other scientific instruments, but tilting platforms cost extra. And, for some, it would be a zero-sum game. Diverting the mission to investigate Michele's anomaly would result in scientists working on some of the other instruments, missing out on taking measurements that were of interest to them. Persuading them that a squiggle in the magnetic field data was more worthy of investigation than the data they had been dreaming about gathering for a decade was not the easiest of tasks. One person's 'must see' can be another's 'waste of time'.

'It was quite a difficult thing to do,' Michele admitted. But she prevailed. Enthusing about the possibility that 'there might be something there', she got her way. And the team spent the next two and a half months working out what would need to be done to accommodate this change of plan.

For several nights before the third, now much closer flyby, Michele 'didn't sleep very well'. 'If we hadn't seen anything at all, the chances are that no one would have believed anything I said ever again,' she said.

'And that was possible?' Jim said.

'It was entirely possible.'

Cassini went within 173km of the surface, for logistical reasons approaching Enceladus from the south pole. The magnetometer was poised, ready to gather data. No one anticipated what happened next. *Cassini* flew directly into a hot, dense jet of water vapour spurting out of the surface of the moon, like lava shooting violently out of a volcano.

No wonder the magnetic field data had looked a bit unusual. A plume like this explained the extra signature in the data from the second flyby, and secured Michele's reputation as someone whose advice was worth taking and whose professional opinion could be trusted.

On later visits, *Cassini* went to within 25km of the surface of Enceladus, risking everything for more and better data. The scientific return, however, was to die for. Organic material was detected on the surface of Enceladus and in the plume. And measurements of the gravitational pull near the surface suggested significant amounts of water were present. A small pond had been assumed. Now a liquid-water ocean seemed more likely.

'Boom!' Jim said, excited. 'You had liquid water, organic material and an internal source of energy. Why do I not recall the media going absolutely crackers over this discovery?'

'The *Cassini* scientists were very careful about how

they described their results.' And the results had come in gradually. 'We didn't discover everything at the same time,' Michele explained. There was a slow build-up of evidence, not a sudden revelation, as is so often the case in science. Scientific papers on Enceladus were published piecemeal in academic journals. With each new flyby, more evidence emerged suggesting that conditions on Enceladus might once have been conducive to sustaining life: another necessary, but not sufficient, condition was met. No single discovery provided definitive proof. There was no obvious tipping point. 'And we've kept our heads,' Michele said. 'We haven't played the life card.' Evidence of life 'might be there but we still don't know, twelve years later'.

'And I suppose if you play the life card too early, before you know definitively, then you don't win anything?' Jim said.

'Right. I always make sure never to use the "life" word, because what we're trying to do as scientists is work out whether a place has the potential for habitability. That's the word I use.'

Journalists can get very excited, sometimes overly so, about the search for alien life, and some of the scientists involved in the search can experience feelings of excitement too. 'I think they sometimes hype it up themselves,' Michele said. 'One of the things that scientists always worry about is getting funding for the science they do, and if you can put the life card on your science, the chances are it will look sexier than if you can't.' Even the money people are not immune to the intoxicating possibility of finding life elsewhere, however old, tiny and small. 'I always try to filter out that kind of stuff,' Michele said. Nevertheless, she is clear about what finding liquid water and organic material on Enceladus means. 'From my point of view,' she said, 'the potential for life of some kind out there is pretty high. But we can't just go out there and find it. We have to understand the planetary system before we can see whether life is possible.'

'I always make sure never to use the "life" word, because what we're trying to do as scientists is work out whether a place has the potential for habitability. That's the word I use'

Jim interviewed Michele eight months before the *Cassini* mission to Saturn performed its final suicidal dive into the rings of Saturn in September 2017, bringing the mission to an end 20 years after it began. (Initially planned to last just four years, it was extended several times.)

'We're going to have twenty very close orbits, we're skimming the cloud tops. Right at the end of the mission, as we dive into the atmosphere, we're going to try and measure the internal planetary magnetic field . . . To do that we need to get as close as we can.'

'It does sound like it's going to be pretty spectacular,' Jim said.

'It is,' Michele said. 'And I'm a little unsettled by it all because what we're trying to do with both the spacecraft and the instrument, neither were designed to do. I worry about my instrument because it's what I have control over. The other thing that is worrying people is that we are going to run out of fuel before the end.'

'Yes, I can see that would be a problem,' Jim said wryly.

'Jokingly, the person responsible for operating the

spacecraft said to me, "We'll probably be on the fumes by the time we get there."'

'We talk about Saturn as being a gas giant, but we don't know whether it has a solid centre, do we?' Jim said.

'We're almost sure, right at the centre, there's a solid core. And then there's a liquid region above it. And we think it's probably metallic hydrogen, because you need there to be current flowing to generate a magnetic field,' Michele explained. 'So it's going to be really exciting, but I'm beginning to lose sleep over it, I must confess.'

'Presumably this is a once-in-a-lifetime opportunity?'

'Absolutely. I doubt we will be able to do this again in my niece's lifetime.'

'And will you, when it's finally over, shed a tear?' Jim asked.

'Yes. Because it's been twenty years of my life!' Michele exclaimed. 'The plan is that most of us will go out to JPL to watch it actually end. I think it happens about three o'clock in the morning, local time. We'll then probably all have a good sob, go and get a few hours' sleep, and have a great party the next day to celebrate everything we've done together.'

When Michele was unexpectedly promoted to become a Principal Investigator on *Cassini*, she was an uncertain leader. 'For a long time on *Cassini*, I pretended I knew what I was doing. After a few years, I think I'd convinced everyone. And after even longer, I convinced myself that I knew what I was doing.'

These days she has the confidence to say, 'Oh, I don't know the answer to that. I'll go away and find out.' And, now that she has turned 50, she finds 'I don't really care what anyone thinks. If I think it's the right path, I do it anyway.'

'I think all of us over fifty feel like that,' Jim said. 'Isn't it great!'

'I just wish I could have been like that in my twenties,' Michele replied. 'But what the hell!'

Barcroft Media/Getty Images

EUGENIA CHENG

'If you never feel like you've pushed your brain to its limits, it will never stretch'

Grew up in: Sussex
Home life: partner (a singer), parents, sister, brother-in-law, and two nephews
Occupation: mathematician and concert pianist
Job title: Scientist in Residence at the School of Art Institute of Chicago
Inspiration: her mother Brenda, who introduced her to mathematical ideas that made her brain contort out of its skull
Passion: the mathematics of mathematics
Mission: to rid the world of maths phobia
Advice to young mathematicians: confusion is part of understanding – you can't make progress without it
Date of broadcast: 23 January 2018

Nothing annoys Eugenia Cheng more than the suggestion that there is no creativity in mathematics. She doesn't spend her time multiplying big numbers in her head. She sits in hotel bars drawing (mainly arrows) with a fine-nibbed artist's pen, thinking about how ideas from different areas of mathematics relate to one another. A relentless desire to get to the bottom of things has driven her to ever higher levels of abstraction, rejecting first physics, then applied maths, then algebra for being too superficial; before specialising in Category Theory, an area of mathematics that seeks to reveal an underlying logic that unites the whole of mathematics. Determined to encourage us all to think more mathematically about the world, she left her job as a mathematics lecturer at the University of Sheffield to pursue a portfolio career and is currently Scientist in Residence at the School of Art Institute of Chicago, where she teaches mathematics to artists, many of whom would rather avoid the subject.

Keen to communicate mathematical ideas to the widest possible audience, Eugenia Cheng has written a book of mathematical recipes and appeared on primetime American TV, armed with a rolling pin, to show how the French patisserie *millefeuille* is made and demonstrate the power of exponential growth. A square of pastry is folded over on itself from either side to create three layers and then rolled out flat to its original size. Repeat this process just six times and you have *millefeuille*, a thousand leaves of wafer-thin pastry – or 2,187 leaves, to be precise ($3 \times 3 \times 3 \times 3 \times 3 \times 3$, or 3^6).

'What I like about your *Life Scientific*, Eugenia, is the way you have managed to combine all your different passions,' Jim said.

'Music, maths and baking are things I've been very keen on since I was very small,' Eugenia said. Her parents, having moved to rural Sussex from Hong Kong, were keen to help their children to assimilate and introduced Eugenia and her older sister Alethea to 'all the things that British children seemed to do'. While many of their Chinese friends were busy learning Mandarin, they practised the piano and made cupcakes, like their neighbours.

'Music, maths and baking are things I've been very keen on since I was very small'

'Baking is about taking basic ingredients and putting them together to make something delicious, and maths is like that too.' The trouble is, 'maths doesn't always come across like that in lessons.' Too often mathematical ideas are introduced as a set of rules that 'you just have to follow', when in fact the fun begins when mathematicians make up their own recipes. 'You start seeing what you can create and then seeing whether you like it or not,' Eugenia said, breaking into a big smile.

'Were you always good at maths?' Jim asked.

'I definitely always loved it,' Eugenia said. 'And I was lucky to have a mother who is mathematical. She didn't say, "OK, now we're going to sit down and do maths. OK, now we're going to do baking."' Maths was 'just there all the time'. Mathematical discussions were part of life.

One of Eugenia's earliest mathematical memories is of sitting in her favourite green armchair at home, listening to her mother talking about how you can make a drawing of the square numbers on a graph. As she described how, when you plot square numbers on a graph, you create a curve that gets steeper and steeper, forming a parabola, Eugenia remembers 'just being amazed that an idea could turn into a picture'.

'The chair is not very big but I remember it as enormous, so I must have been very small,' she said. 'My brain just felt like it was contorting itself out of my skull.' She wanted to understand how such a thing was possible, and she has relished the feeling of 'her brain stretching out of her skull' ever since. Physical challenges have never particularly appealed to her, but she enjoys flexing her brain muscles to make them stronger and likes to remind her students that if their brains are hurting, then 'it's because they're getting cleverer'.

'I think sometimes people don't like maths because they think, "Oh, my brain is getting very confused, I must not be very good at this" – but actually it's because that's what maths is like.' Confusion is part of the path to mathematical enlightenment. Being bewildered and befuddled is inevitable if you eventually want to understand. 'You can sit there for one, two or twenty years without really getting anywhere,' Eugenia said.

'You can sit there for one, two or twenty years without really getting anywhere'

Playing the piano delivers more immediate and more reliable rewards: 'you practise and you get better.' As soon as her older sister started having music lessons, Eugenia, aged three, would listen in, annoyed that she had to wait until she was five before she could begin, a habit that continued even when she started having lessons of her own. Sometimes she would surprise her teacher. 'Did I tell you that?' her teacher would ask. And, quick as a flash and pleased as punch, Eugenia would shout out: 'No. I heard you tell my sister!'

The only thing Eugenia and her older sister ever fought over was whose turn it was to practise. As soon as Eugenia had mastered one piece that was too hard for her (and sometimes before), she would be given a new one to learn. 'And so, we kept progressing,' she said. She gave solo recitals when she was very young and won her first international piano competition aged 11.

'Did you practise a lot?' Jim asked.

'From the age of five, I practised really a lot,' Eugenia said. 'But not because my mother made me,' she said, articulating every word loudly and clearly to make sure we got the message. 'People often say that small racist thing of how you must have had a Tiger Mother, and my mother gets very offended because she really let us do whatever we wanted.'

Eugenia's un-tiger mother worked as a statistician in the City of London, and Eugenia remembers meeting her at the station every day, a rare woman in a sea of male commuters. Her father, 'a real people person', worked for the NHS as a child psychiatrist, caring for some of the most vulnerable members of society. Seeing her suited mother on the platform, briefcase in hand, surrounded by similarly clad men, didn't strike her as particularly notable at the time. It didn't worry her that, unlike all her friends' mothers (who stayed at home), her mother never picked her up from school. And, reflecting on the experience now, she thinks it was very important. 'I had this baseline assumption growing up that I could do anything a man could do,' she said. It helped protect her from gender stereotyping herself.

For a long time, she didn't think her situation was unusual, but looking back, she realises how privileged she was. 'I now see that many girls don't grow up thinking that,' she said. 'Indeed, many boys don't grow up thinking girls can do anything they can do. It was only later I realised that there was this terrible stereotype that maths is more for men. And when I did get to Cambridge it was very, very male-dominated.'

'How difficult was that for you?' Jim asked.

'I don't remember it being difficult at the time.' She had been well prepared by the Director of Studies at Roedean School, who had warned her: 'When you get to Cambridge there will be a whole load of boys who will all be better than you. They will have been pushed really hard at school.'

'So, when I got there and I wasn't the actual worst, I was actually quite pleasantly surprised,' she said.

As her teacher had predicted, the boys were way ahead.

'They just breezed through, boasting about how easy it all was,' and learning how to deal with their arrogance 'and be OK with it' was a challenge. Later, however, the tables turned. When some of these self-assured boys were doing PhDs, 'they had forgotten how to work hard,' Eugenia said. 'They couldn't hack it and didn't want to do it any more.' Eugenia meantime went from strength to strength, enjoying the mental exercise, relishing the feeling of her brain stretching out of her skull, just as she had done in the green armchair at home when she was very young. 'A lot of people who had appeared to be better than me fell by the wayside during PhDs or during their first post-doc jobs, and I just carried on.'

By the time Eugenia was doing her PhD, she had specialised in Category Theory, the mathematics of mathematics, the purest of the pure. It's a branch of mathematics so abstract that some fellow mathematicians think it goes too far.

'What drove you to such a high level of abstraction?' Jim asked.

'I really was that two-year-old who never stops asking why,' Eugenia replied. 'And I was never satisfied until the answer became really logical and fundamental.'

When she was doing her GCSEs she thought it would 'be really nice' if she could just do maths and physics. At A Level she found physics unrewarding. 'Formulae would come out of nowhere and no one ever explained why they were there.' When she read mathematics at Cambridge, only pure maths appealed to her. Mathematical physics, for example, was too applied. She was not interested in describing the nature of the physical universe. She was searching for a purer, more abstract and more fundamental truth. Algebra was, for her, 'the bit of maths that best explained how things interacted' and her master's degree was algebra-only. But then she found herself wondering how algebraic interactions worked.

'So finally, I started my PhD and I was doing only Category Theory, and then I thought . . .' Eugenia said, laughing, 'wouldn't it be nice if I could just do higher-dimensional Category Theory? That, to me, is the pinnacle.'

'Of abstraction, of purity?' Jim suggested.

'Well, yes, because each of these stages explains the previous stage,' Eugenia said.

Higher-dimensional Category Theory is the theory of Category Theory. And the wonderful thing is, when you reach for higher dimensions and enter this dizzying abstract world, 'the theory loops back on itself' and brings closure. Even the most dogged desire to know why is satisfied.

'The theory is itself!' Eugenia exclaimed with great enthusiasm. 'And that's why higher-dimensional Category Theory is the pinnacle.'

Category Theory does for mathematics what mathematics does for science. John von Neumann supplied the mathematics that Albert Einstein needed to get to grips with relativity. (He worked down the hall from Einstein at Princeton's Institute for Advanced Study.) The mathematician Roger Penrose and the theoretical physicist Stephen Hawking worked together in the 1960s, spending many hours side by side, poring over large sheets of paper that they had laid out on the floor to help them think about what happens in the middle of black holes. The mathematics developed by David Hilbert in 1909 to study infinite dimensional space was used about a decade later to formalise quantum mechanics. Often, however – and rather sadly for many mathematicians – the usefulness of their particular piece of mathematics is only realised long after they have died.

Mathematical logic is timeless and has been used by science through the ages. It provides ready-made sets of tools and rules that save scientists from having to work everything out from scratch every time they encounter a new situation in the real world. If the scientists can find some off-the-peg mathematics that will fit, there is no need for anything tailor-made.

'Maths explains how science works,' Eugenia said. 'It takes that part of science that all scientific subjects have in common with each other and says: "Well, let's do that bit separately, so that scientists can focus on the specifics."' It reveals how the same logic can govern scientific systems that might be wildly different. The same mathematics, invented by Carl Friedrich Gauss in 1809, can be used to study the distribution of grains of sand, or IQ scores. Both are described by a bell-shaped curve, with an equal number of measurements above and below the mean value.

Similarly, Category Theory identifies the underlying principles that govern disparate areas of mathematics and hopes to reveal a unifying logic that applies to them all. What mathematical way of thinking remains constant, regardless of the context? What is the mathematical logic that will work across the board?

'A lot of the power of maths comes from making connections between things,' Eugenia said. 'So I think it's a very unifying subject. And it's not just about getting the right answers, it can be about illuminating a situation and finding ways to think about it more clearly.' It's also about being efficient. It's a way of avoiding unnecessary repetition. 'I sometimes think of it as front-loading all the effort. You invest in a situation so that you never have to make any effort in that area ever again.' Or put another way: 'Good maths comes out of being really lazy.'

'Could you give an example of what Category Theory is about?' Jim asked.

'One of the driving principles behind it is to say that we can understand a lot about things by their relationships with other things, rather than their intrinsic characteristics,' Eugenia said. 'A philosopher might ask: "What is the number one?" A category theorist would ask: "What does the number one do? What role does it play in relation to other numbers?"'

To which the answer is: when you multiply any number by the number one, the number remains unchanged. When you add one, a number increases, but the unique property of the number one is that when you multiply with it, nothing happens. A category theorist would then search for analogous behaviour in a different area of mathematics – in the world of shape, for example. He or she might ask, when you multiply shapes together, which is the shape that performs the same role as the number one?

When you multiply a circle by a line, you get a cylinder. Imagine stretching out a slinky. When you multiply a circle by circle, you wave a circle shape in the shape of a circle. If you extend a slinky so that it describes a circle, joining the ends together, then you get a shape like the inner tube of a bicycle, or a bagel. But when you multiply a circle by a point, nothing happens. The circle remains in one place, the point. Its shape does not change; it remains a circle. And so, a category theorist would say, a point and the number one are essentially the same. Intrinsically they may be very different things but they relate to other objects in their respective worlds of shape and number in the same way.

Jim wondered how Eugenia was able to capture such abstract ideas: 'What does the world look like when you're doing your maths?'

'There's not a lot to see,' Eugenia said, smiling. 'Mathematics happens in the inner workings of the brain, which is not something we can draw or see in normal life.'

When Eugenia is doing mathematical research, she sits very still, moving things about in her head. She used to work in cafes, but when cafes started feeling more like libraries, she started going to bars. Hotel bars, in particular. The flow of people makes her feel like 'there's a lot of possibility'. Sitting quietly with an artist's notebook and an espresso, she feels like a 'little calm point' and hopes to create something inside her brain.

Creating imagery in her brain, however, is only half the battle. 'At some point you have to translate it onto the page.' She has to find a way to tell people what she has seen. To aid this process, she likes to use a fine-nibbed artist's pen because 'it just feels sort of more special and free.'

'Are there days when you sit staring at a blank piece of paper and nothing comes?' Jim asked.

Sometimes. But she always makes herself do something. She sits down for an hour a day, without fail. 'It's amazing how much progress I can make like that,' she said, and then reminded us of some basic maths. 'If you add up lots of ones you will eventually get a very large number.' By trial and error, she discovered, this strict daily schedule is the only way she can cope. 'I'm holding all these pieces in my head,' she said. 'If I don't go back to them every day they just get messed up. If I stop for just two days, it all starts to go wrong. If I stop for a week, it's hopeless.' She ends up with no idea what she was thinking about a week ago. So she just keeps going 'in little increments' and if she feels stuck, she tells herself to fill a page, regardless of whether or not she thinks the content is any good.

'How do you visualise higher-dimensional ideas? What does that look like on the page?' Jim asked.

'If you show a mathematician a diagram with lots of arrows, they'll say: "That looks like Category Theory,"' Eugenia said. She draws pictures to represent abstract ideas and arrows to indicate the relationships between them. Her mathematical drawings are designed to reveal underlying truths.

The map of the London Underground provides a good analogy. The tube map is not geographically accurate. The Circle Line doesn't follow a rectangular path. Lancaster Gate is a lot closer to Paddington station than it appears. But it doesn't matter. The map has been carefully designed to reveal the connections between stations, not the reality on the ground, or even underground. Each new graphic designer who has worked on the map of the London Underground has aimed to make the underlying structure of the system both clearer and more aesthetically appealing (as well as adding new stations and line extensions where necessary). The aim being to enable us all to access the information we need with minimal effort and create a map that looks good too.

Eugenia similarly spends a lot of time thinking about how she can organise her diagrams so that they make 'instant sense'. 'I'm quite visual like that,' she says. 'I love moving diagrams around so that they look better.' And she does this because she wants to encourage an emotional response. The most informative tube map is also the most stylish. She wants her mathematical diagrams to look good, and she would seem to be succeeding. The mathematical drawings in one of her notebooks were displayed as a work of art in the Royal Academy Summer Exhibition in London in 2004.

'To be really successful at thinking logically about a problem, we have to have an emotional engagement with it'

'We think of maths as being super-super-logical,' Eugenia said. 'But to be really successful at thinking logically about a problem, we have to have an emotional engagement with it.'

As Eugenia went deeper and deeper into Category Theory, reaching for ever higher level of abstraction, she engaged both her intellect and her intuition, attributing the former to her 'searingly logical' mother and the latter to her instinctive father. When abstract ideas are drawn well, it helps us to access our intuition to solve a problem, be it planning the most efficient route between stations or tackling higher-dimensional Category Theory.

Having specialised in such an abstract area of mathematics, employment opportunities were few and far between. A series of post-docs in Cambridge, Nice and Chicago were 'fun but stressful': as soon as she got one job she was applying for another and Eugenia lived in constant fear of imminent unemployment, no matter how well she was doing. She expected that moving to America would be tough. It wasn't. Chicago, she was delighted to discover, was full of musical opportunities. She gave solo recitals and started accompanying singers performing in the annual Schubertiade and at other musical events organised by Pianoforte Chicago. Playing the piano was a reassuring counterpoint to the vagaries of mathematical research: she practised and she got better. And it gave her the most visceral pleasure.

Rather unexpectedly, however, she experienced culture shock when she returned home to the UK to take up a permanent job as a mathematics lecturer at the University of Sheffield. 'The whole education system had been . . .' She paused for a moment. 'How shall I put it?' She paused again. 'Brutalised? Is that too strong a word?'

'You can say what you like, Eugenia,' Jim said.

'It had certainly changed since I was last there,' Eugenia said.

From day one in Sheffield, she got stuck into teaching. It was something she had always done. When she was doing her PhD, she taught A Level students. Being mathematically able herself, it seemed to make sense to focus on the most able pupils and help to prepare them well not just for A Levels but also for university. But she soon found herself thinking: 'If only what they did last year had been done differently,' and she moved herself down a year to fix the problem. This pattern repeated as she taught progressively younger children, until she concluded that the most useful contribution she could make to the school system was to teach mathematics to five-year-olds.

She was also interested in helping schools to develop their maths-teaching strategies. Aware that many schools needed (understandably and rightly, in her view) to focus their resources on helping children who were struggling, she thought one way in which a university could and probably should help was to provide mathematically able children with the opportunity to be challenged. In this way, she hoped, more children from local schools could arrive at university brimming with mathematical confidence and be comfortably able to hold their own.

She loved teaching students, inside and outside the university, but found it a terrible struggle to find any time to do her own research. 'There was so much administration and so much bureaucracy. At the beginning of every academic year, I felt as if I was not really going to breathe again until after the summer resit exams in August,' she said, 'leaving just two or three weeks to frantically get some research done. You barely have time to remember what on earth it was that you were doing a year ago, before it all starts up again.'

Every September she found herself thinking: 'Ohhh, it's another year. How did that happen?' Until one year she thought: 'Gosh. Is this how it's going to be until I retire?' She had also been finding it difficult to get her voice heard within the university. 'I didn't know what was going on but I started to feel as if I wasn't having as much effect as I wanted.'

For the first few years, Eugenia assumed this was because she was still junior, but then newer, younger lecturers started to arrive and seemed to command more respect than she did. 'They also all happened to be "maler" and whiter than me,' she said.

'Was that part of your decision to leave?' Jim asked.

'Yes, it was. Definitely.'

'Still, that's a very hard decision?'

'It's definitely unusual for anyone to give up a permanent position,' Eugenia agreed.

She began to think that maybe she could be more effective operating outside the system rather than within it. To help her decide what to do next, she turned to Category

Theory, firmly believing that this approach to problem solving did not need to be confined to mathematics and could be applied to other areas of life. 'If you want to be the biggest fish in the pond,' she reasoned, 'either you can grow or you can move to a smaller pond.' And, according to Category Theory, one way to make the pond smaller is to investigate more characteristics. Her calculation worked like this. By carefully defining the area in which she operated as being the set that contained all of her talents, not just a few, she increased the probability of there being fewer fish.

She knew plenty of other mathematicians who, like her, were trying to rid the world of maths phobia, but how many of them were female and not white? 'I thought to myself, I'm not the best mathematician in the world. And I'm not the best public speaker in the world. But maybe I could be the best mathematician public speaker in the world.'

Having been brought up and educated to believe she had a duty to make the best use of all her talents in order to make a positive contribution to the world, she sat down and made a list of everything she was good at. And then she thought: 'If I stop teaching undergraduates these courses, someone else will do it, but who is helping to get rid of the maths phobia in the world? And who is combating the unfortunate stereotypes about mathematicians?'

'There's so much in the popular media and in films of mathematicians who are older, white guys who are a bit strange,' Eugenia said. She was fed up of seeing people on screen who 'couldn't make eye contact and then, excitingly, turn out to be really good at maths'. She wanted to challenge the idea that 'to be good at mathematics you have to be a bit weird and have no friends'.

In an attempt to break down barriers between the two cultures, many scientific institutions now have artists in residence. Far fewer art institutions invite scientists to move in, but in 2014 Eugenia became the Scientist in Residence at the School of Art Institute of Chicago, responsible for teaching mathematics to artists. 'To develop a complete mind,' said Leonardo da Vinci, 'study the science of art; study the art of science. Learn how to see. Realise that everything connects to everything else.'

Few of the painters, sculptors, photographers or graphic designers Eugenia teaches, however, enjoyed maths at school. All of them were put off by the same things: 'memorising, especially times tables, formulae, equations, right and wrong answers, rules that come out of nowhere'. So she makes sure to do 'none of that'. Instead, Eugenia exploits the exploratory nature of mathematics. She wants her students to experience feelings that are 'like wandering through a jungle and seeing what beasts you can see poking their heads out of the undergrowth.'

She gives her students very hands-on projects. Sometimes they try to build something. Other times 'there's no particular aim', the idea is simply to explore the mathematical terrain. 'I give them things to do whereby no matter what they do, they will discover something,' she said. Building structures out of tetrahedra, for example. The individual tetrahedrons are perfectly symmetrical. (The four triangular faces are identical and interchangeable.) What happens when they start putting them together? She asks her students to describe their work: is what they have created symmetrical? Or interesting in other ways? Does it please them, and if so why? Are the mathematical constructs they have created delicious? Or disgusting? By letting her students do what they most enjoy, she builds up their emotional engagement with mathematics while they play.

'Would you say you are trying to get these artists to enjoy mathematical logic in the same way they might enjoy listening to Bach?' Jim asked.

'Yes. Maths, unfortunately, is something where people often think, "Oh, I'm no good at it, so I'm not going to interact with it any more." But it is possible to appreciate mathematical thinking even if you can't do it yourself . . . you can enjoy going to an art museum even if you can't paint. I think maths can be like that too,' Eugenia said.

She knows her art students are unlikely to become mathematicians. 'I'm not trying to turn everyone in the world into category theorists,' she said with a smile. But she does want to help people appreciate maths, and the first step is to get people to understand why maths exists, 'what it's here for'. If students then start using mathematical logic to think about their own lives differently, 'that's even better'. And when students tell her that studying mathematics has affected their art practice, she is delighted.

'I would love to stop seeing articles in the media saying things like, "I've never had to use a quadratic equation in my life."' Mathematics is a phenomenally powerful tool. It underpins a lot of science. It helps us to think clearly. But, Eugenia believes, to judge mathematics solely on its ability to be useful is to miss the point. Mathematics is creative. It irks her that so many people think otherwise. 'I don't blame those people,' she said. 'We mathematicians are clearly doing a very bad job of conveying what maths is really about.'

She tells her students all they need to take a course in mathematics is 'to be curious and open-minded'. She wants to teach them different ways to think about the world around us. And that, she says, is what maths is about.

'And baking delicious cakes along the way,' said Jim.

LUCIE GREEN

'I do love the fact that the sun is a contained nuclear bomb'

Grew up in: Cranfield, Bedfordshire
Home life: married to maths communicator Matt Parker
Occupation: astrophysicist
Job title: Professor of Physics at the Mullard Space Science Laboratory, UCL Department of Space and Climate Physics
Inspiration: seeing beautiful high-resolution images of the sun through a telescope in Crimea
Passion: the sun
Mission: to understand coronal mass ejections
Best moment: asking Buzz Aldrin if astronauts see flashes of light in their eyes as high-energy cosmic particles whizz through space in front of them
Most worried about: the threat to Earth from extreme space weather events
Advice to young women scientists: 'Don't let men do all the talking'
Date of broadcast: 3 October 2017

Lucie Green studies coronal mass ejections. She wants to know what causes jets of high-energy plasma to escape from the surface of the sun, typically releasing as much energy as a billion hydrogen bombs and with the potential to create electrical chaos here on Earth. A major coronal mass ejection hit our planet in 1859, wrecking telegraph networks in Europe and America. A smaller one hit Quebec in 1989, leaving 6 million people without light or heating in their homes. And there was a near-miss just before the London Olympics in 2012. Had this coronal mass ejection arrived a week later, it's estimated that 10 per cent of the satellites in orbit around the Earth would have been taken out. The most likely net result for many of us: no TV, no mobile phones, no internet.

'Did you always want to be a scientist?' Jim asked.

'No,' Lucie replied firmly. 'As a teenager I was absolutely set on being an artist.' Initially, art school was glorious. Lucie loved being free to express herself and enjoyed the physicality of creating art, but she soon started to yearn for some logical structure. She had been 'endlessly enthusiastic about physics' at school and missed having a 'right answer to work towards'.

'Art is open to subjectivity in a way that physics isn't.' There were no rules, not even guiding principles. One conversation with her art tutor about a piece of work she'd done was particularly distressing. She knew what she was doing. She had thought it through. And then the tutor came and said, 'Oh, I can see what you are trying to convey here,' and Lucie thought, 'You know what, that isn't what I was trying to convey at all!' Such a cavalier disregard for truth felt deeply uncomfortable. 'I couldn't do that kind of subjectivity for a living,' she said. 'So that brought me back to physics.'

Aware that, as Galileo said, the language of the universe is mathematical, and, more pragmatically, knowing that few university physics departments would accept her without maths A Level, she studied mathematics in her spare time and sat the exam while she was finishing her art foundation course. And she read and read about the physics of the stars, becoming ever more entranced. 'By the time I got to Sussex University, I was absolutely desperate to learn more about the universe.'

She visited the Crimea Astrophysical Observatory as an undergraduate. Close up, the sun is 'beautiful and structured. It has these lovely loops of hot flowing gas in its atmosphere and it changes over time.' It was both visually and scientifically exciting: a winning combination for Lucie. 'It's the ultimate scientific object,' she said, and 'aesthetically interesting'.

Jim, a nuclear physicist, had a rather different view of the sun: a giant, violent star at the centre of our solar system capable of generating nuclear power on an astonishing scale. Pairs of hydrogen atoms fuse together to form helium, releasing 400 trillion, trillion watts of energy every second. Or, put another way, the energy produced by the sun in a single second is enough to power civilisation as we know it for nearly half a million years. 'It's basically a giant nuclear fusion reactor,' Jim said.

'I do love the fact that the sun is basically a nuclear bomb,' Lucie said. Temperatures inside the sun reach several million degrees kelvin. Luckily, cooler material nearer the surface contains this giant ball of hot gas and prevents the whole thing from exploding.

'I do love the fact that the sun is basically a nuclear bomb'

Seeing her first solar eclipse in 2009 was a defining moment in Lucie's relationship with the sun. 'There's something very intriguing about seeing another side to our local star,' she said. She had seen images of the sun before, but they had all arrived via a computer screen. They were facsimiles, not the real thing. And she was excited to be able to look at the sun's atmosphere unmediated.

Photons that had been formed deep inside the sun, released when two hydrogen atoms had fused to form a single heavier atom of helium, were entering her eyes.

'There's this popular conception that sunlight is eight minutes old,' Lucie said. 'It's not. It's hundreds of thousands of years old!' The sunlight that's in the air right now was formed hundreds of thousands of years ago in the middle of the sun when *Homo sapiens* first started to appear on planet Earth. 'I love the fact that we're seeing something that is that old,' Lucie said.

'Photons made in the very heart of the sun have a hard time getting out.' These particles of light interact with the electrically charged material of the sun itself and 'sort of ping around like a ball in a pinball machine'. Hundreds of thousands of years later, they escape from this giant ball of hot, charged gas and emerge into the vacuum of space. From there, the journey is straightforward: 93 million miles in just eight minutes.

Aware that job opportunities for people with a keen interest in the sun were few and far between, Lucie applied to do a PhD at the Mullard Space Science Laboratory in Surrey in 1998. When her dad wondered what she might do if the PhD idea didn't work out, she realised there was no plan B. 'Everything was focused on the sun.'

The day of the interview arrived. She was picked up from Guildford station and driven to MSSL, 'going down progressively smaller and smaller country roads'. Important signs of civilisation soon disappeared. 'Where is the pub?' Lucie thought. 'Where is the bus stop?'

When they started to drive up a single track, she thought, 'I can't possibly come and work here.' She was living in Brighton at the time and enjoying the urban, cosmopolitan vibe. 'There was really nothing there,' she said.

She answered every question in the interview that followed, utterly convinced that she didn't want the job. There was no way she could spend three years in such an isolated and lifeless place. A few hours later, however, she changed her mind.

She was taken on a tour of the MSSL, starting with an exhibition in the main house where pictures of British-made rockets hung on the wood-panelled walls. Surprised to learn that Britain had been so actively involved in the early days of space exploration, she read on. In the 1950s Harrie Massey and Robert Boyd, two physics professors at University College London, were thinking about how to get into space and finding their central London offices a little cramped. They had both worked on projects during the Second World War in historic buildings, away from town centres and university control, and, inspired by how scientists and engineers had worked together towards a common goal, hoped to recreate a similar working environment for their rocket science. They found a magnificent Victorian mansion in the middle of the Surrey countryside and set up the Mullard Space Science Laboratory as a joint science and engineering facility.

It worked. In 1957, just months after the Soviets sent *Sputnik* into orbit, the Mullard space scientists launched *Skylark*, a rocket designed to gather data on the upper atmosphere and briefly experience conditions in space. It was the first British rocket to go into space, albeit only for a few seconds, and between 1962 and 1969 UK scientists and engineers (many of whom owed everything they knew to the ongoing British *Skylark* rocket programme) worked with NASA to launch and build satellites. *Prospero*, the first British-made satellite to be launched on a British rocket, went into orbit in October 1971. It's still up there.

Lucie was shocked. How could she not have known? And how exotic: a group of scientists and engineers, the pioneers of space science, working away, hidden from the world, in this splendid country house.

The next stop was the lab, where she watched modern-day MSSL engineers making instruments and subjecting them to brutal tests to see if they could survive in space, and talked to scientists who could barely contain their enthusiasm for the work they did. Pub or no pub, she concluded, MSSL was 'the best place in the world'. She got through the interview, accepted the job and has worked at MSSL ever since.

'We are one big family,' Lucie said. 'Because we are alone in the countryside, we focus very much on working together.' Ideas fly between scientists and engineers: it's Harrie Massey's dream come true. 'The scientists ask questions,' Lucie said. They need hardware to get the data they need to answer those questions. 'More questions get asked. More instruments are needed. And you have a virtuous circle.'

'And the swimming pool and the croquet lawn are a bonus?' Jim asked with a smile.

'The really nice thing about being surrounded by engineers 'is that we can fix everything we have on site. So if the swimming-pool pump breaks, we can fix it.'

In her student days, colleagues fixed things without waiting to be asked. On more than one occasion Lucie has found her bicycle with a Post-it note attached, describing the work that has been done. When you spend all day designing things that are fit for the utterly hostile conditions in space, earthbound challenges are trivial. And a pleasure to pursue.

The engineers were good party people too, often rigging up flashing LED lighting or borrowing solid carbon dioxide to make dry ice to make social events more exciting.

When not partying, Lucie studied coronal mass ejections. 'We are all living in the atmosphere of the sun,' she explained. The heliosphere, an immense magnetic bubble, embraces our entire solar system, extending well beyond Pluto, and it determines the geomagnetic conditions in space. When this 'space weather' interacts with the Earth's atmosphere, we get the Northern Lights. But not all such interactions are as pretty or as benign.

Coronal mass ejections (or giant solar belches, as Lucie likes to call them) are the most energetic events in our solar system, releasing about as much energy as a billion hydrogen bombs. A billion or so tonnes of highly charged particles escape from the surface of the sun and speed through space. And when these jets exit the sun from a particular point and at a particular angle, it can have serious consequences here on Earth.

They had been known about for just over 20 years when Lucie started her PhD in 1998, but no one knew how or why they happened. 'I was interested in what it is that enables them to become eruptive,' Lucie said. 'How do you lift the mass of a mountain off the surface of the sun, where the pull of gravity is 27 times greater than what we experience here on Earth?' Moving mountains on Earth is hard enough without solar gravity making things 27 times harder. Clearly, some serious forces must be at work. 'We knew it had something to do with the sun's magnetic field, but we didn't know what,' Lucie said. She wanted to work out exactly how the magnetic forces in and on the sun contributed to a coronal mass ejection: what are the magnetic structures that cause this violent and massive outpouring of solar matter?

The magnetic field lines in and around the sun are constantly changing, becoming twisted, convoluted and distorted. Nothing is stable. (Forget neat arrangements of iron filings around a horseshoe magnet; the sun is a maelstrom of magnetic activity.) 'You can see the sun as a spherical ball of electrically charged fluid,' Lucie explained. 'And a special consequence of this is that the magnetic fields that thread through the sun are "frozen in". So as the fluid moves through the sun, it drags the magnetic field with it.'

It's violent and chaotic and very dynamic. This moving magnetic field (created by circulating electrical fluid) injects energy into the system, energy which can then be unleashed in a coronal mass ejection. They can happen several times a day with varying degrees of intensity. Bubbles of spluttering matter burst, and jets of high-speed plasma spurt out from assorted locations on the surface, hurtling through space in every possible direction, at hundreds of thousands of kilometres per second.

Viewed from Earth, the only evidence of these enormously energetic events are solar flares, patches of the sun that seem to shine more brightly than usual. (From a distance of nearly 100 million miles, appearances can be deceptive.) But these solar flares were enough to give the amateur British astronomer Richard Carrington a clue. He was the first person to suggest there might be a connection between the bright spots on the surface of the sun and some dramatic events and geomagnetic storms here on Earth, having spent more than seven years staring at the sun from his private observatory in Redhill, Surrey. On 1 September 1859 he first observed a dramatic solar flare. It coincided, he noted, with some extreme readings on the magnetometer in Kew Gardens and dramatic events here on Earth the following day.

The resulting aurora was so bright that 'the light

appeared to cover the whole firmament', according to reports in the *Baltimore American* and the *Commercial Advertiser*. Gold miners in the Rocky Mountains got up in the middle of the night to start work, imagining it was morning. Pylons started to spark. Entire telegraph networks all over Europe and America went down. Strange happenings indeed, caused by a coronal mass ejection that came to be known as the Carrington Event.

In March 1989, 6 million people in Quebec found themselves in the dark and without heating as the temperatures outside dropped to minus 15 degrees Celsius. Subway trains sat still, traffic lights didn't work, and petrol pumps no longer delivered fuel. Questions were asked of the local energy company, but a coronal mass ejection was responsible, not a corporation. Four days earlier, a solar flare had released a giant speeding jet of hot-charged plasma. It hit the Earth's atmosphere and spiralled down the North Pole, distorting the Earth's magnetic field and causing electrical chaos.

In July 2012, just as London was preparing to host the Olympics, a jet of charged particles passed through the Earth's orbital path just weeks ahead of planet Earth arriving at the very same place. It was a near-miss. Measurements of the magnetic-field strength, particle density and plasma speed all pointed to a coronal mass ejection of Carrington proportions. Had it arrived a few weeks later, we would all be aware of the disruptive power of a coronal mass ejection.

It was bad enough in 1859 when the telegraph network was the height of electro-magnetic sophistication, a cutting-edge technology. Today, thousands of satellites orbit the Earth, unprotected by the Earth's atmosphere. Most of our communication and navigation systems depend on them. GPS, mobile phones and the internet all rely on satellite technology, as do military surveillance systems, the stock market and the National Grid. Not to mention television coverage of the Olympics and accurate clocks to time the 100 metres.

Carrington-style events are thought to be a one-in-eighty-year, or one-in-a-hundred-year event but, as Lucie warned, 'it can happen any time'. And the consequences would be more serious than they were in 1859. 'Space is so much part of our lives' and the satellites on which so much of modern life depends would be sitting ducks.

If a Carrington-style ejection were to visit us today, it is estimated that 10 per cent of all our satellite fleet would be lost, according to a report by the Royal Academy of Engineering. GPS alone is thought to underpin 6 per cent of the GDP of the developed world. The global cost of an extreme space weather event could be more than a trillion dollars.

'Do you see coronal mass ejections coming?' Jim asked. 'It takes a few days to get to us from the sun, so presumably we have some warning that it could be heading in our direction?'

'For the really fast coronal mass ejections, you don't have very long.'

The Carrington coronal mass ejection took just 17 hours to travel 93 million miles from the sun to Earth. If it were possible to predict the location and timing of the next solar flare, then maybe we could create an early warning system for events like these and so know days, not hours, in advance that a coronal mass ejection was hurtling towards planet Earth. Electrical power systems could be shut down and emergency plans put in place.

'But that's not something we are able to do yet?' Jim asked.

'We need to know the physics of coronal mass ejections. Why they happen. We need the right models to forecast them. I think we're probably a few years off that yet'

'Not yet!' Lucie said. 'That's the holy grail of space weather reporting. At least it is for me, but it's a huge amount of work . . . We need to know the physics of coronal mass ejections. Why they happen. We need the right models to forecast them. I think we're probably a few years off that yet.'

'Around 2011 solar activity was picking up' (it waxes and wanes in eleven-year cycles) and Lucie started to try and raise awareness of how space weather could seriously disrupt modern life. She talked about coronal mass ejections to anyone who would listen and made a dramatic video about the consequences of an extreme space weather event hitting planet Earth that was shown at science festivals and distributed on social media.

'We scientists often feel as if we are banging our heads against a brick wall, trying to get governments to listen to our concerns,' Jim said. 'How easy was it to get government ministers to take space weather seriously?'

'It has been a slow burner,' Lucie said.

Solar scientists were fully aware of the potential hazards in the year 2000, but 'it wasn't deemed a significant risk'. The new science minister, David Willets, however, was a great ally. Lucie met him at a science festival when they participated in a panel discussion together. 'David was very interested in space science and very knowledgeable,' she said. 'He understood the risks that space weather presented' and helped to persuade the British government to take it seriously, as did the UK Space Agency, which was created in 1985 to co-ordinate British space science.

'I think these two things were the key.'

'Was it the term "space weather" that captured the imagination?' Jim asked.

'I suspect the fact that the *Daily Mail* kept producing headlines about how the sun was going to kill all life on Earth didn't hurt!'

Headlines that declared 'Freak space storm caused by solar changes could scupper London Olympics' added to the drama. But perhaps the real secret of the scientists' success was to convert the risk into cash, a tactic employed by the economist Nicholas Stern to get us all to pay attention to climate change, in the *Stern Report* of 2006. Money talks.

'The economic risk would be significant, and I think it was about conveying that to government,' she said. 'We spent a lot of time trying to actually work out what the economic impact would be . . . Once you can start to talk about that, if it's in orders of magnitude, then you can start to persuade people that it's worth putting in a very small fraction of that money to mitigate the risks.'

It would seem to have been a success. In January 2012, geomagnetic solar storms were added to the National Risk Register alongside pandemic flu and terrorist attacks. The UK Met Office now monitors space weather 24/7, having opened a dedicated space-weather operations centre in 2014. It now issues regular forecasts describing the level of disturbance for three types of space weather: radio blackouts, geomagnetic storms and solar radiation storms. Keen to avoid their pigeons getting confused, pigeon racers take an interest, as well as key industries and government authorities. (Pigeons use the Earth's magnetic field to navigate and can be thrown off course during a heavy solar storm.) 'Perhaps one day the space weather report will appear after the shipping news on Radio 4,' Jim said, jokingly.

Lucie has designed several measuring instruments for the first-ever dedicated space weather satellite, which is about to be built by the European Space Agency and due to launch in 2023. Just as satellites in orbit around the Earth gather data on atmospheric conditions that are used to improve terrestrial weather forecasting models, so this new satellite will gather more data on electromagnetic conditions in the heliosphere, which will be used to improve our understanding of space weather.

If you put a satellite at a special point between the Earth and the sun, known as the 5th Lagrange point, it will maintain a steady orbit between the Earth and the sun, trailing just behind the Earth. 'You're sort of looking around the sun' to see what's coming next. It also means the spacecraft can collect data along the sun–Earth line, the slice of space weather that matters most to us.

Another way to get to grips with space weather is to get to know the sun better. Between 2009 and 2011 Lucie was the chief observer on the Extreme Ultra-Violet Imaging Spectrometer instrument (on the *Hinode Satellite*, a Japanese Aerospace Exploration Agency project with US and UK collaboration). It was put into sun-synchronous orbit around the Earth, at an altitude of about 600km, and has been sending back non-stop footage of the surface of the sun since 2006, supplying us all with some spectacular images of our local star, documenting eruptions and solar flares. And generating more data on coronal mass ejections.

Since then she has been busy designing instruments for the European Space Agency's forthcoming mission to the sun. The *Solar Orbiter* is due to launch in October 2018, orbit the sun Icarus-style and 'look at this furnace from close up'. It will be exposed to sunlight 13 times more intense than what we experience on Earth. Temperatures on the sun-facing side of the spacecraft will reach 500 degrees Celsius and will drop dramatically when the spacecraft's elliptical orbit takes it much further away.

'It's a serious thermal challenge,' Lucie said. It's hoped that a coating of a newly created heat-resistant material, Solar Black, will provide 'a very significant heat shield' to protect instruments that operate best at room temperature, not these solar extremes. And 'the tiniest windows have been put in' for the cameras to 'sort of peek through'. 'I love the fact that I can go down the stairs, through the common room to the labs and see the kit,' Lucie said, enthused. 'I can't touch the instruments any more. I'm not allowed to do that.' The engineers have had too much experience over the years of curious scientists, concerned about the birth of their invention, inadvertently making unhelpful interventions. 'But I can see it being made.'

Nearly twenty years after she first wandered around the Mullard Space Science Laboratory and changed her mind about not wanting to work in such an isolated place, Lucie's enthusiasm for the science and engineering work done in the grounds of a splendid Victorian mansion in the Surrey hills has not waned.

Tiina Särkinen

SANDY KNAPP

'The more I find out, the more I realise I don't really know'

Grew up in: Los Alamos, USA
Home life: three children and ex-husband Jim
Occupation: botanist
Job title: Head of Plants, Natural History Museum, London
Inspiration: identifying plants in the desert in New Mexico
Passion: the *Solanum* genus of flowering plants
Mission: to catalogue and understand the diversity of the *Solanaceae* family of flowering plants
Best moment: 'Seeing something new. It happens all the time'
Worst moment: having giardia in Peru
Advice to young scientists: 'Talk to people who are not like you'
Date of broadcast: 24 January 2014

Sandy Knapp is passionate about plants. An unstoppable force when hunting for specimens, she has travelled extensively in Central and South America and has spent a lot of time in Peru. She had wanted to study cacti but was persuaded to look at the *Solanum* genus, a highly diverse and economically important group of flowering plants that includes tomatoes, potatoes and aubergines. Now that she is a world expert on *Solanum*, she is grateful for the introduction. In 2004, she launched the Solanaceae Source Project as part of the Planetary Biodiversity Inventories, an ambitious initiative to try to describe all the different species that exist on Earth. It took Sandy five years to describe all 1,400 species in the *Solanum* genus, and as soon as she finished she started work on the wider *Solanaceae* family.

Sandy grew up in the desert, immersed in 'a sea of physicists'. No one in Los Alamos ever asked, 'What does your father do?' The answer was obvious. Instead they wanted to know: 'Can you talk about it?' Her father designed and built particle accelerators (not atomic bombs), collaborating with Soviet scientists throughout the Cold War, and her parents were regular visitors to the Soviet Union, often staying for a month or more. When they entertained Soviet scientists at home, Sandy and her brothers would enjoy their favourite game: hiding behind a large potted plant on the way to the sitting room, trying to spot the KGB spy.

'Presumably the big burly chap in the corner?' Jim suggested.

'Usually the one whose eyes glazed over when they started to talk about physics,' Sandy replied.

Los Alamos was 'an amazing intellectual environment'. Many of her teachers, even when she was very young, were the wives of PhD physicists who themselves had PhDs (often in physics). Prevented from working in the same laboratory as their spouses by strict nepotism laws, they got jobs in local schools, and all the Los Alamos children received an excellent education as a result.

Living in this physics-saturated desert town, however, exerted an equal and opposite force on Sandy. She used to argue with her father, telling him physics was 'a terrible world': 'You have only eight things [the fundamental particles], I have millions.'

'One of the great things about a liberal arts education is that you are not canalised into a particular thing early on, but I thought marine biology would be really fun,' she said. Having lived in New Mexico and later landlocked Switzerland, the idea of spending a lot of time at the beach appealed, and marine life seemed more exciting than inanimate matter. 'But the course was full, so I decided to take botany instead.'

It turned out well. Most of her time at Pomona College in California was spent in the desert, collecting plants. Her parents used to joke that she was majoring in fieldwork, not botany. 'We went out every weekend with microscopes,' she said, carrying cases which were the size of a microwave oven and almost as heavy. The edges would 'fold up to make a little field table and we'd sit there on the ground with a dissecting microscope and identify plants in the desert . . . It was amazing. Absolutely amazing.'

Sandy's tremendous drive and energy filled the studio. Her passion for plants was inescapable, leaving Jim a little bemused. 'For someone like me, whose research involves equations and computer codes, what is the buzz?'

'Finding out new things,' said Sandy. 'The more I find out, the more I realise how little I really know, which means it reinforces itself endlessly. Around every corner you find something you've never seen before. Every plant you see, every bird . . . is something never seen before in that circumstance. I've seen condors in the Andes before. But this last trip we were on, I saw one that was as big as a cow. As big as the cattle on the slope below. It's just incredible!'

'Around every corner you find something you've never seen before. Every plant you see, every bird . . . is something never seen before in that circumstance'

After graduating, Sandy

wanted to continue her study of desert plants, but her PhD supervisor at Cornell University persuaded her to embrace the *Solanum* genus, an eclectic group of flowering plants that includes several economically important crops. Cacti, not edible crops, were Sandy's thing, but having fallen in love with taxonomy and keen to continue naming and classifying plants, she agreed and wrote her PhD on an obscure group of shrubby *Solanum* species.

They do have some fantastic Latin names, Jim pointed out: '*Solanum tuberosum* [potato] sounds like a Harry Potter magic spell.' Not to mention *Solanum lycopersicum* (tomatoes) and their wild cousins, *Solanum pimpinellifolium*.

Midway through her PhD, 'which seemed to be going on forever', an opportunity to go plant hunting in Panama came about, and she grabbed it. The Missouri Botanical Gardens had embarked on a Flora of Panama project and urgently needed more specimens. She was given a trailer to live in and a truck to take out collecting, and was told to 'Go collect!'

A year later she returned to the USA, loaded with plants and speaking fluent Spanish. She finished her PhD and then decided to go to Peru. Her husband Jim, a lepidopterist, was chasing the passionflower butterfly and Sandy wanted to be with him. She also needed something to do while she was there, so before she left she wrote to everyone she knew: 'I'm going to Peru for a while. Who will pay me to collect plants?' The response was disappointing. Many people said, 'We'll give you a dollar for every identified plant' – 'which is ridiculous. I mean, honestly!' But the Director of Missouri Botanical Gardens was generous. He asked Sandy how much money she wanted. Then doubled it. And with $8,000 in her pocket, off she went.

Sandy and her husband Jim arrived to find Peru in the middle of an armed conflict between Maoist guerilla group, the Shining Path, and the Peruvian government. For a while they lived near the stadium in Lima. 'I can remember waking up in the middle of the night hearing gunshots and thinking, "Oh God, I hope it's not like Chile!"' General Pinochet shot hundreds of innocent civilians in the stadium in Santiago in 1973. She once went down the main road and saw dead dogs hanging from lamp-posts. But most of the time they were in the countryside on the other side of the Andes, and while Jim chased butterflies, Sandy hunted for plants.

They stayed for a year, but when Sandy was seven months pregnant with her first child, it was time to leave. 'I understand you had some trouble getting out,' Jim said. 'Tell me what happened.'

'Ah . . . so . . . well, that was exciting,' said Sandy. 'I had nearly lost the baby actually, in the forest. The hospital near to where we were living had no running water, so I thought we'd better go back to Lima. That was fine. He was fine.' They then decided to go back to the UK to have the baby. Just as she was leaving the country, Sandy was stopped by a young policeman at Lima airport.

'You'll take off all your clothes!' he shouted, shoving a machine gun into her pregnant belly.

'No, I won't!' she shouted back. 'Under no circumstances will I take my clothes off for you. Get a female officer!'

Forced to strip naked by a female officer, Sandy tried in vain to explain her situation. Meantime, her unborn baby kicked the female officer's hand from the inside out and got the message across. Only then was she satisfied that a baby, not a stash of cocaine, was responsible for Sandy's bump.

'They were just looking for drug smugglers,' Sandy said. 'Fair enough. But I think they could have been more polite.'

Alfred was born two months later in the Whittington Hospital in London. From there they went back to the USA. Isabel and Victor were born in Mississippi, within two years of each other. It's not easy for two scientists who are married to each pursue their passion. Three children, born within a six-year period, complicate things further. Often academic couples moan about 'the two-body problem', so named after a scenario in physics in which the motion of a pair of particles that interact with each other needs to be determined, but in London Sandy and Jim found a solution. He joined University College London and Sandy got a job at the Natural History Museum. Fieldwork was more or less put on hold while the children were growing up, and instead Sandy set about cataloguing the plants of Mexico and Central America, compiling an inventory of 18,000 species. *Flora Mesoamericana* was the first taxonomical reference book to be written in Spanish, an idea which had seemed obvious to Sandy. Surely you would want the people in whose country the plants grow to be able to read and make use of this valuable resource?

Carl Linnaeus, the founding father of taxonomy, gave organisms carefully considered and methodically constructed two-part Latin names: the genus name followed by the species. He named us *Homo sapiens*, for example: members of the *Homo* genus, the wise (*sapiens*) species with high foreheads and thin skull bones. The *Homo* genus is part of the wider family of hominids which, together with apes and monkeys, are classified as primates, the order of mammals with collar bones and grasping fingers.

Linnaeus' *Systema Naturae*, first published in 1735, provided a framework for thinking about all the organisms that were known at the time, and it is still very much in use today: a universal system that everyone can follow. 'Latin names may seem a bit arcane,' Sandy said, 'but actually they keep things in order and keep things right.' Taxonomy brings order to the chaos. It's a way of organising organisms, a filing system for the natural world.

Naming and classifying species helps us to think more clearly about all the different plants and animals on the planet. (Latest estimates suggest there could be as many as 100 million different species, with the vast majority of them as yet undiscovered.) 'It's hard to care about something that doesn't have a name', and taxonomy is 'the first step on the road to conservation'. It helps us to appreciate the incredible diversity of life and makes it easier to think about how all the different species are related. Linnaeus believed all the different plants and animals were the product of God's creation, representatives of an unchanging order. A century later Charles Darwin and Alfred Russel Wallace introduced the idea of certain all species sharing a common ancestor and of their change through time by evolution by natural selection.

'It's hard to care about something that doesn't have a name'

Taxonomy has 'got three legs, like a stool', Sandy explained to Jim. One leg describes the diversity of life, another leg reveals the evolutionary relationships between different species, and the third leg makes all this information accessible to as many people as possible. 'Without one of those legs, it doesn't stand up properly. It doesn't really work.' Species have to be accurately described and catalogued and filed according to the same rules. (For organisms, as for paperwork, it is not always obvious what belongs where.)

The naming system invented by Linnaeus has been added to and improved steadily over hundreds of years. At the start of the twenty-first century, however, many botanists feared their skills and knowledge would be sidelined. Compared with the new science of genomics, Linnaean taxonomy seemed out of date. The Human Genome Project was launched to great fanfare in June 2000 by UK Prime Minister Tony Blair and US President Bill Clinton, and the genome of the model plant *Arabidopsis thaliana* (a biologically useful weed) was sequenced in the same year. Work had begun on rice (sequenced in 2002), grapes, papaya, sorghum, cucumber and maize, to name but a few. And in 2003, a consortium of plant geneticists decided to tackle the tomato.[1] There was no shortage of money for scientists who sequenced genomes, it seemed, and it was easy for botanists to feel resentful. But rather than panic or grumble, Sandy spotted an opportunity. She repackaged her botanical skills and presented them as something that could be of service to plant genomics.

'I convinced my colleagues in the US that we needed to put together a project to look at *Solanum*,' she said. 'And much to our amazement, we got the grant from the National Science Foundation to do a global monographic revision.' They even got some money from The Plant Genome Project, which felt like a real coup. Securing funding to conduct a thorough review of the characteristics

1 The tomato genome took nine years to complete.

of all the members of her favourite genus was a dream come true. It was also a great excuse to go back to Peru. When Linnaeus created the *Solanum* genus, it contained 23 species. Now there are more than 1,400, including all sorts of different species of tomato, many of which grow in the wild in Peru, hanging from cliffs, busting through the canopy of the cloud forest and growing on the side of the road. *Solanum pimpinellifolium* has tiny red fruits the size of raisins and thrives in coastal regions. *Solanum chmielewski* grows in the high, dry valleys of the Andes. When crossed with the cultivated tomato, the fruit from the resultant plants are exceptionally sweet and much loved by the ketchup industry for this reason.

Once fears had been allayed that this attractive-looking fruit was not poisonous (despite belonging to a plant with leaves that looked suspiciously similar to those of deadly nightshade), humans started to cultivate different varieties of tomatoes to deliver delicious and resilient fruit. Banana legs, black plums and chocolate stripy are just a few examples of the more exotically named varieties. Forget yellow, orange, green and red – we now have tomatoes with skins that are cream, crimson and deep purple. The Gigantomo has fruits the size of footballs. Humans have created thousands of varieties, which have caused considerable taxonomical confusion. The fruits may look very different, but they all belong to a single cultivated species, *Solanum lycopersicum*.

The Solanaceae Source Project, created by Sandy and her team, is a comprehensive online resource which will ultimately contain digital images of all the 1,400 species in the *Solanum* genus, and includes details on their habitats and pollinators, and the herbicides that destroy them. Interactive identification keys help plant collectors classify their finds, and there are links to relevant research in plant genomics, alongside more traditional botanical descriptions.

'One of the stipulations of our grant was that we had to make connections between our world and their world,' she said. Between botany and genetics. 'People think all biologists are the same but we're really, really different!' she exclaimed. 'We inhabit totally different planets.'

Geneticists can happily have no idea what the plants they study look like. They spend days in the laboratory with pipettes, preparing DNA samples, looking down microscopes at cells.

'I was the only one who was really interested in this whole collaboration thing,' Sandy said. 'I thought it sounded fun. Therein began my career talking to genome scientists . . . The plant geneticists have made me look at my world in a different way.' And she hopes she has done the same for them. 'And maybe this could be a model for looking at how to integrate across the biological sciences. Setting biology in a diversity context makes it richer.' It's important to keep an eye on the big picture.

'We've brought biodiversity and genomics a little bit closer together,' said Sandy, 'at least for the *Solanaceae* family.' The Solanum Project was an ongoing project designed to be constantly updated, but just when it was looking as if the bulk of the work was complete, Sandy decided to extend this already ambitious project to include all the species in the wider *Solanaceae* family, opening the door to petunias, tobacco and capsicum peppers, and thousands more species of nightshade vegetables and fruits.

Sandy's enthusiasm for plants seems limitless. She described her work as a constant voyage of discovery. 'It's clear to me that your love of plants has kept you passionate about your work for all these years,' Jim said. 'Is it really that simple?'

'Nothing is ever as simple as it seems,' Sandy said. 'I've been immensely lucky in the people who've hired me and places I've been able to go. But I've also tried really hard to have a good time.' Being a scientist is, she believes, 'a licence to have a good time'. It's the perfect excuse to explore and find out new things.

'It is,' she concluded, 'the most creative and wonderful thing you can do with your life.'

Imperial College London

SANJEEV GUPTA

'We took the sea away from the English Channel'

Grew up in: Dehradun, India until the age of five, and then Sevenoaks, Kent, and Reading
Home life: married to consultant haemotologist Dr Anja Drebes with two children, Tarak and Uma
Occupation: geologist
Job title: Professor of Earth Sciences at Imperial College, London
Inspiration: hiking in the Himalayas
Passion: understanding how landscapes have formed
Mission: 'I don't have one. I randomly do things with a gut feel'
Advice to young scientists: 'Follow up intriguing ideas and chase them'
Date of broadcast: 14 May 2013

Sanjeev Gupta takes the longest possible view of history. Inspired by a trek in the Himalayas and looking for an excuse to visit more spectacular locations, he decided to study geology and has studied mountain ranges and rock formations all over the world, on the bottom of the English Channel and on Mars. In 2007 he proved that Britain's separation from the European continent half a million years ago was caused by a catastrophic megaflood, showing that is was a dramatic event and not a case of the two land masses gradually drifting apart, as had previously been thought. Encouraged by a colleague to apply to NASA to work on the Mars Science Laboratory mission, he became head of long-term planning, responsible for deciding where the Martian rover *Curiosity* should go next. Jim interviewed Sanjeev in March 2013, a month after *Curiosity* had made history by drilling into the surface of Mars to extract a rock sample, and before the results had been analysed.

Sanjeev's colleagues often make jokes about him being like Dylan, the rather dopey guitar-playing rabbit in *The Magic Roundabout*.

'Do you recognise the likeness to Dylan?' Jim asked.

'Umm. Some of it. Yes,' Sanjeev said.

'So unflappable, calm, chilled?'

'Chilled, mostly,' he said. 'I like to sit on a hillside with a thermos flask, making sketches and thinking.'

Sitting outside, sipping tea and enjoying the view, he spots clues that are hidden in the landscape and buried with the rocks, and thinks about how they might have moved over geological time to create what he sees before him. 'You can see features that you can't explain,' he said. 'Then, in a sudden flash of inspiration, you see a narrative that could bring together all your seemingly random observations into one coherent story. You can't worry about being wrong. If your hypothesis doesn't work out, you just have to be able to develop another one. Thinking quickly in the field.'

'You can't worry about being wrong. If your hypothesis doesn't work out, you just have to be able to develop another one'

According to Sanjeev, 'All Indian parents want their children to be doctors,' but it was the chance to travel to exotic destinations that excited him, not medicine. At school, he loved history 'because of the nuances of interpretation', and he didn't find science particularly inspiring. 'It all seemed to be completely sort of set in stone . . . There was no mystery to it, in the way that it was taught. There wasn't any way you could put something into it.' Scientific knowledge appeared too fully formed. It didn't seem to want him to think for himself.

He spent his childhood in Sevenoaks and Reading, his parents having moved from Dehradun in India when he was five. When he was 17, he decided to return home to visit his family in the foothills of the Himalayas and 'took the chance to go trekking'.

'High up on the edges of the Tibetan plateau is this just extraordinary landscape, with its deep ravines and gorges, and everywhere rocks of different colours, shapes and varieties.' Sanjeev did wonder, briefly, how the landscape had come to be the way it was (as you might expect from a future geologist): why was it so very different from the gentle hills in Kent? But the thought that really motivated him was: how could he get a job that could take him to places like this, 'because this is fantastic, I love this'?

Hoping to maximise his chances of visiting more spectacular landscapes, he went to Oxford University to read geology and soon started planning an expedition to Borneo, thinly disguised as a geology field trip to study some little-documented coastal rocks. He wrote to the British army's jungle warfare training school, who were stationed nearby, and they agreed to help. And so it was that four scientists, one British officer, three Eban tribesmen and four Gurkhas made their way up the Belait river, having been taken as far as possible by an army helicopter. They stayed in the last longhouse on the river, sleeping on the ground and looking at faded portraits of the Sultan of Brunei and the Queen that hung on the mud walls next to a series of Japanese skulls in baskets, 'macabre relics of World War Two'.

'I remember Sanjeev being very relaxed, as always,' said his friend Bill McCaffery, chief teller of the Dylan jokes. 'He didn't flap very easily.'

'It sounds more like a *Boy's Own* adventure than a geology field trip,' Jim said.

'Oh yes, it was!' Sanjeev replied enthusiastically. 'You absolutely want to run into the rainforest and play, but you can learn a lot from that, you know. I think we knew we weren't going to do anything dramatic scientifically, but it was about exploration and putting yourself on the edge. It was hard. It was tough.'

Hungry for more adventure after graduation, he spent time in the remote Chinese province of Chengdu, having convinced a Chinese research institute that his geological services would be useful. (He offered to make a map of the rocks in an area that hadn't been geo-mapped before.) To prepare for the trip, he taught himself Chinese and organised all the necessary logistics, which was no mean feat. His parents, meantime, worried about his tendency to dream.

Most of his PhD was spent sitting on the sides of mountains in the Alps with his thermos, dissecting the view in his mind and thinking about the geological history of the region. Later, he went to the deserts in New Mexico and Sinai, where he remembers looking into the red desert rocks and thinking, 'These rocks were formed under the sea – that's remarkable!' Geology had given him a 'time machine'. It also brought him back to what he had so loved about history at school, 'the nuances of interpretation'. Geology is history, albeit on a much longer timescale than is used by most historians. Evidence is pieced together (from a variety of rocks, not people) and different stories are told that try to make sense of it all.

'The best theories are the wackiest ones,' Sanjeev said. 'You're trying to look for clues to see if a theory holds up.' When claims are sufficiently grandiose, 'it's often something you can test. With something that's less spectacular', often you can't tell the difference between one hypothesis and another. The rocky outcomes would be broadly similar in both scenarios. 'But if it's something really outrageous, you can actually say yes or no.'

Sitting in a library one day, he came across an old paper on the origins of the English Channel. The author, Alec Smith, was a little-known academic at Bedford College in London whose contribution to science seemed largely to have been ignored. Most geologists at the time assumed that Britain detached itself from the Continent gradually. The received wisdom was this: about 450,000 years ago, various rivers (in what is now the English Channel) burst their banks, flooding assorted alluvial plains, and over time these ill-defined rivers and pools gradually coalesced to form a body of water that separated England from France.

Alec Smith, however, envisaged a much more dramatic series of events. His paper (published in 1985) suggested 'A Catastrophic origin for the paleo-valley system of the eastern English Channel'. Imagine the white cliffs of Dover joining up with similar chalky structures in Dieppe to form a continuous structure known as a rock ridge. To the north-east, behind this chalk ridge, is a giant lake enclosed by what is now the coast of East Anglia, Holland and Denmark. The chalk-face has been sealed by a permafrost, so no water can get through. To the south-west, land runs continuously from southern England to northern France. Early humans could walk from Dover to Calais. Then picture this: ice sheets come down from the north and melt in this giant lake. The chalk barrier can't take the pressure of all this extra water and the chalk dam bursts. A torrent of water rushes through the broken rock ridge, flooding the dry land to the south-west, and, almost overnight, Britain becomes an island.

Not many people had read Alec Smith's paper. Fewer still had taken it seriously. It was just 'too wacko', Sanjeev said. And it didn't have enough evidence to back it up. But its radical nature appealed to Sanjeev. True or false, at least he would get a definitive answer. And, if it were true, how intoxicating! Here was an opportunity for Sanjeev to prove something – including, perhaps, himself.

Sadly, the people who gave out research grants were less convinced. 'Some of the most interesting things are hard to get funding for,' he said. 'But I have to be passionate about the cause.' Gripped by a desire to understand the origins of the English Channel, he applied to ever more obscure sources of funding and eventually succeeded.

The plan was to use the latest sonar technology to survey the bottom of the English Channel. By bouncing a pulse of ultrasound off the sea floor and timing how long it takes to return, depth measurements can be taken at any number of locations. In this way Sanjeev hoped to 'remove the water from the English Channel' and get a sense of the topography of this underwater landscape.

'What I wanted to do,' Sanjeev said, 'was to go to the Straits of Dover and look at any cataclysmic features that might be there.' Evidence of dramatic events ought to be fairly obvious and would support Alec Smith's dam-busting hypothesis.

But Sanjeev's colleague Jenny Colliers spotted a flaw. 'There's no way we're going there!' she exclaimed when Sanjeev told her of his plans. Taking state-of-the-art sonar technology into the world's busiest shipping lanes did not feel like a good idea to her. More than 500 ships pass through this narrow stretch of water daily.

Jenny's common-sense approach prevailed, and they started mapping a small, less busy area of the Channel. They were unsure at this stage what else they could usefully do. 'Then what happened was one of those wonderful coincidences in science.' The UK Hydrographic Office (the government department in charge of mapping the sea floor to help prevent ships from running aground) spotted what Sanjeev and Jenny were doing and were curious. They wanted to know more about the smart new sonar equipment they were using.

Sanjeev and Jenny happily showed off their new kit and explained how it worked, and the two teams started to collaborate. The UK Hydrographic Office had been mapping the bottom of the English Channel for 30 years and all the data had been stored in their archives.

'We could suddenly see the landscape for the first time. So in the centre of the Channel you could see these beautiful valleys carved into the bedrock where, in all likelihood, early humans would have walked and hunted'

Using data that was 30 years old but new to science, Sanjeev generated images of the bottom of the Channel. 'When we saw those first images it was just an astonishing moment,' he said. 'It was one of those jaw-dropping moments. We took the sea away from the English Channel. We could suddenly see the landscape for the first time.

So in the centre of the Channel you could see these beautiful valleys carved into the bedrock where, in all likelihood, early humans would have walked and hunted.'

The valley in the centre of the Channel was much bigger than anyone had thought. The images also revealed a set of distinctive features that were utterly unfamiliar to Sanjeev: elongated island-like humps with beautiful streamlined shapes and grooves that ran for many, many kilometres along the sea floor.

It's tempting to believe that scientific knowledge has advanced because we've become increasingly clever; but Sanjeev believes: 'We just have better ways of seeing things in more detail.' Progress in science is driven largely by technology. It's also about making connections. Sitting in the library once again, 'leafing through a journal looking for something else', he came across a drawing of a feature that had been found on the surface of Mars. It made him pause for thought because it reminded him of the strange humps he'd identified at the bottom of the Channel.

'I'd actually never shown any interest whatsoever in Mars before this,' he said. But the resemblance was striking. An excited rush of research followed as Sanjeev tried to find out more about these Martian features, hoping they might shed some light on similar features here on Earth. The Martian humps were, he found out, associated with very vigorous water flows. Giant canyons nearby were also thought to be caused by a tumultuous Martian flood. All of a sudden, Alec Smith's radical hypothesis was starting to look a little less ridiculous.

For three years, Sanjeev gathered evidence in support of Alec Smith's idea, reading around the subject and endlessly analysing the data. Like a barrister preparing for a difficult case, he paid close attention to every detail of this dramatic geological story, then spent ages imagining ever more obscure counter-arguments and preparing robust responses.

His paper entitled 'Catastrophic flooding origin of shelf valley systems in the English Channel' was published in 2007, endorsing Alec Smith's dramatic 1985 hypothesis and confirming Britain's sudden exit from the Continent. Despite the radical, violent and unusual nature of the separation suggested, most scientists were convinced by Sanjeev's evidence (and still are). And the scientific consensus on how Britain became an island was overturned.

Case closed, you might think, but Sanjeev had no intention of resting on his laurels. 'We have the first set of evidence and we have low-resolution data. What we are working towards now is trying to build a better set of pictures. That's the scientific process,' he said, cheerfully accepting that there was an enormous amount of work still to do. And suggesting, perhaps, that Sanjeev hasn't given up on the idea of one day mapping the sea floor in the Straits of Dover with state-of-the-art sonar technology, however busy the shipping.

Soon after he'd sorted out the English Channel, a chance encounter propelled Sanjeev into space. He was presenting a poster at the National Space Centre in London, showing how features found on Mars had helped him to solve the mystery of the English Channel. Jan Peter, an engineer from the Mullard Space Sciences Laboratory, had the spot next to him and was showing off the high-resolution stereo camera that he had built for the *Mars Express* mission that had been launched in 2003 and was still sending back data.

They got talking and went for lunch together and before long were happily collaborating. Jan Peter had been creating 3D models of a Martian landscape that were much clearer than anything Sanjeev had managed to recreate. Pictures of the bottom of the English Channel, which is never more than 174m from the surface, were blurrier than those taken by the *Mars Express* spacecraft of a surface 93 million miles from Earth.

'It wasn't difficult to persuade Sanjeev to leave Earth behind, at least temporarily,' Jan Peter said, 'because he was absolutely fascinated by the pictures and models we'd created of the Martian surface.'

'Suddenly there was this technology and this engineering knowledge to enable us to see better and see a new place for the first time,' said Sanjeev. 'And it was really exciting, Jan Peter showing me these three-dimensional views of Mars.'

'So, you decided to start studying Mars,' Jim said. 'Had you run out of unexplored places on Earth?'

'No. There were plenty of unexplored places, but I just couldn't go to them at that stage. We had young children and my wife was very busy. She's a doctor in a London hospital and it was very difficult to go off to these exciting, exotic places like the Andes or Tibet.'

Grounded by the geological adventures he had enjoyed as a young man, Sanjeev was unsure what to do with his career, but after giving it some

'I thought . . . I could study Mars from my kitchen over a glass of wine'

thought he came up with a plan. 'I thought, well, actually, I could study Mars from my kitchen over a glass of wine.'

Having made the decision to move to Mars, he started to prepare for this shift in focus. 'You can't enter a new field as an interloper without knowing something about it,' he said. 'There's nothing that upsets people more than when you launch in with your theories and you know nothing about what's happened.' For a year before he started working with Jan Peter, he did nothing but read – catching up on all the literature about Mars takes time.

The collaboration worked well. 'Jan Peter knew how to create these images and I knew what they meant,' Sanjeev said. They were working on an in-depth study of what looked like an ancient delta in the Eberswalde Crater, rebuilding high-resolution data sets to get a more precise picture of conditions on Mars millions of years ago.

The Eberswalde delta, meanwhile, was rising up the popularity charts as a possible landing site for the latest Mars rover, *Curiosity*. 'Unbeknown to me,' Sanjeev said, 'there was a whole competition going on where scientists were bidding for their favourite landing site to be selected.' The Mars Science Laboratory mission was due to launch in 2011, carrying the state-of-the-art rover, and opinion was deeply divided about where it should land. So NASA had set up a competitive process and invited the public to vote.

Different groups of scientists favoured the particular patches on the Martian surface about which they already knew the most and were not so keen on the idea of *Curiosity* landing anywhere else. You could say it was 'gloves off'. The Eberswalde lobby recruited Sanjeev to their cause and he went to a few NASA meetings and made the case for *Curiosity* landing in the Eberswalde Crater. It would be sheltered and quiet, he said and, since it showed all the signs of being a delta, it seemed like a sensible place to look for evidence of water.

Eberswalde made it into the final four but no further. 'Unfortunately, your site wasn't chosen in the end,' Jim said. 'Were you disappointed?'

'A little bit,' said Sanjeev. He was also 'a bit relieved because, you know, it's my geology deciding an expensive rover.' He didn't want decisions that could cost NASA a lot of money to rest solely on his professional opinion about the geology of Mars, especially given that it was based on evidence that was sparse. People who are planning space missions would like doubt to be obliterated. 'Of course, when you're sending a rover that costs $2.7 billion, you have to have a bit of certainty about it.'

The winning destination, Gale Crater, is 'an absolutely fantastic site', Sanjeev said. A great succession of rocks are on display (a bit like the stripy sediments in the Grand Canyon). 'Imagine you're in the Grand Canyon, or somewhere with a thick pile of rocks, that enables you to read the environmental record of Mars,' Sanjeev said. 'NASA was interested in getting that long-term record of the evolution of surface conditions on Mars.'

Soon after the vote and encouraged by Jan Peter, Sanjeev applied to work on the *Curiosity* mission and was given a senior position: in charge of long-term planning (one of just two UK scientists in a 400-strong team of mainly Americans). 'And you're an Earth man, not a planetary geologist,' Jim said. 'Were you surprised to make the team?'

'Well, yes,' Sanjeev said, 'because it was a whole new world for me.'

The study of distant planets used to be confined to looking at photographs taken by orbiting spacecraft. No special geological skill is required there. But with Martian rovers capable of taking close-ups of the rocks and picking up specimens and, in *Curiosity*'s case, hopefully drilling down, NASA had realised that people with proper geological expertise were needed. Reading rocks is not straightforward and it helps to have gained some hands-on experience on Earth before attempting to study rocks on Mars. 'We know the nuances,' Sanjeev said. 'We know what things aren't, actually! So we can rule out certain possibilities.'

On 6 August 2012, *Curiosity* began its descent to the Martian surface. All the scientists involved had gathered in one giant room and were watching events unfold on a big screen, knowing that because it took a while for the signal to get from Mars to Earth, everything they were witnessing had actually happened about 20 minutes ago. As they whooped and cheered when the parachute opened without a glitch, they knew that the mission might have already ended in failure, with a crash landing, for example. So it was difficult to celebrate wholeheartedly. But when they saw *Curiosity* touching down safely, 'the room just erupted', Sanjeev said. 'People were in tears.' Scientists who 'were normally mild-mannered' started hugging each other and crying. Many had devoted more than a decade to building some of the instruments on board. 'You don't often see them so emotional.'

After the landing, Sanjeev was based at NASA's Jet

Propulsion Lab in Pasadena. As things turned out, studying Mars was not the relaxed kitchen activity of his dreams. JPL was a long way from home and alcohol-free. *Curiosity* needed to be monitored constantly for the first three months as it trundled across the surface of Mars, sending back data collected by the 17 cameras on board. Days on Mars are longer than those on Earth, so he soon 'got out of sync' and often had to work night shifts to look after *Curiosity* during daylight hours. 'It was extremely difficult,' he said. 'My wife just laughs at me.' Being a doctor, she just gets on with it. 'But there you are, I found that quite hard.'

'One of the mission goals of *Curiosity* is to look for habitable environments. What do you mean by that?' Jim asked.

'We mean ancient rocks that would have been conducive to life forming . . . A habitable environment is one that had water in it, an energy source to drive metabolism and a carbon source to build carbon-based life. And water is the biggest thing.'

It had been known for a long time that there are canyons and valleys on Mars, thanks to pictures taken by orbiting spaceships. *Curiosity* was designed to look for evidence of water. Once that's found, 'then it gives us really strong clues as to where to look for ancient rocks that might have been habitable,' Sanjeev said. 'Because those are the ones that we have to target if we're gonna go the next step.'

Others believed NASA should be more ambitious. According to space writer Stuart Clark, *Curiosity* should have been designed to look for evidence of life itself, to search for organic molecules and ancient micro-fossils. In his view, 'NASA wanted to play it a little more conservative' because it needed some easy wins. 'If they went to the planet saying that they were looking for life on Mars and found nothing; well, in the current financial squeeze, perhaps that would mean that the mission was perceived as a failure,' Stuart said.

Sanjeev defended NASA's slow and steady approach. Finding ancient micro-fossils is 'a tough call' on Earth, let alone on Mars. 'Whenever people publish papers on discovering microbial life in very ancient rocks on Earth, there's huge kerfuffle. Imagine what it's like with Mars.'

Jim wondered if there were other places in our solar system where 'microbial life might be more abundant' – the moons of Jupiter, or Titan, which orbits Saturn. 'Shouldn't we be focusing our attention on them?'

'I think Titan is an amazing place,' Sanjeev replied. 'But it's conjecture.' Working alone as a young man, Sanjeev had happily hared off after bold conjectures. Now he toed the NASA line, aware that a lot of taxpayers' money was at stake. 'For Mars, we have some knowledge and we build on that knowledge. We know that Mars, before 3.5 billion years ago, is likely to have been warm and wet. This is the time that life on Earth was beginning. So, it's a good idea to go to Mars and look for rocks that are that old, because we know that we find evidence of life in similarly aged rocks on Earth.'

Jim wondered what it felt like working on a big-budget, high-pressure mission with lots of media attention.

'I find that quite exciting, actually,' Sanjeev said. 'I love being able to explain to the public what I do . . . The science I do on Mars is not always rocket science. It's actually quite fundamental classical geology, but we're able to say some really exciting things about a new place, and that's wonderful.'

Studying geology on Mars, however, is rather different from doing it here on Earth. For starters, it takes place indoors. All mountain views are mediated by a screen. 'You can't go out there and look at rocks, or gather evidence, so you're looking for small features, like stratified layers in rocks which could have been formed by water,' Sanjeev said. Field work on Earth is often a solo activity; exploring Mars is highly collaborative. There are lots of scientists involved and 'everyone brings their own unique experiences and knowledge'. A big plus for Sanjeev is that they can, and do, 'discuss crazy ideas about how landscapes formed'.

A few months into the mission (before anyone really expected to discover anything), *Curiosity* found some small, rounded pebbles at Yellowknife Bay in Gale Crater. Here, for the first time, was evidence of fast-moving water that was probably about knee-deep and must have flowed for at least a thousand years. Broken-up rocks don't become rounded and smooth overnight, ruling out the possibility of this ancient valley being formed by a sudden ice melt or a flash flood, for example. The *Daily Express* announced: 'NASA looks set to reveal what could be the most significant discovery of modern times.' It was an overstatement. (Apparently a NASA spokesperson had said it was 'one for the history books' and the journalist had taken things from there.) But it was exciting.

The plan had been for *Curiosity* to drill down into the rocks wherever conditions looked promising, but by the time the data had been analysed and these lovely pebbles had been identified, *Curiosity* had moved on. Five months later, however, the rover returned to Yellowknife Bay to investigate further, drilling 6cm beneath the Martian surface and extracting a grey powder in February 2013. The results of this investigation were published nine months after Sanjeev had been interviewed by Jim. The grey powder was a fine mudstone, containing 20 per cent clay minerals, nodules and veins, all of which are highly suggestive of a wet environment. And for the first time in NASA history these wet conditions were shown to be neither strongly alkaline, nor strongly acidic. Nor were they too salty. The chief scientist on the mission declared that *Curiosity* had found evidence of water that was good enough to drink. It was not direct evidence of life on Mars. (*Curiosity* was not designed to look for that.) But it did suggest 'a habitable environment, so benign'. Yellowknife Bay 3.5 billion years ago was capable of supporting life, albeit in a simple form.

Jim wondered if, having spent several years studying Mars from a distance, Sanjeev might like to visit in person. Denis Tito had recently issued an invitation to two people, probably a married couple (because they would be used to spending long periods of time together), to travel on his private mission to Mars. 'Would you like to go on the first field trip to Mars, and would you take your wife with you?' Jim asked.

'The question should be, would my wife take me?' Sanjeev replied. 'Umm . . . gosh. I'm not sure I'd want to be the first one. The romantic in me would love to go, there's no question about that,' he said. But thinking about the length of the mission, the closed, cramped conditions and the fact that we don't really know very much about our physiological responses to such a situation, he concluded that he would be 'happy for someone else to go'.

'Actually, I really like Earth,' Sanjeev said. 'I find Earth quite exciting.'

'Actually, I really like Earth . . . I find Earth quite exciting'

ACKNOWLEDGEMENTS

The Life Scientific was a brilliant and bold commission by the controller of BBC Radio 4 Gwyneth Williams: a conversation between scientists to be broadcast on prime-time Radio 4, which for reasons I will never understand is 9 a.m. Thank you, Gwyn. This commission is the biggest present I have ever received. The second-biggest present came from Alan Samson at Weidenfeld & Nicolson, who decided to convert these interviews into a book, or six. He bravely agreed to let a radio producer loose in print. Paul Murphy has suffered the consequences of this decision and has skilfully and patiently guided me through this transformation. I am grateful to Paul, and to Holly Harley and Simon Fox, for their impeccable editing skills.

I am overwhelmingly indebted to all the scientists featured in this volume. Thank you for being so generous with your time both in the studio and on the phone beforehand, kindly replying when I said 'just one more question, if I may?' for the seventh time. Thank you all for being so open and so honest. Your stories have inspired me, in many different ways. Your can-do attitude is infectious. Together you have made this book.

A huge thank you is also long overdue to the editor of the BBC Radio Science Unit, Deborah Cohen. I rely on her wisdom and have plundered her encyclopaedic knowledge of science, and scientists, repeatedly. *The Life Scientific* has benefitted greatly from our conversations about science, radio and life in general, as have I. The BBC Radio Science Unit is a unique ecosystem, and I depend on my fellow inhabitants for my survival. Thank you to all my colleagues on the first floor of Old Broadcasting House for their support and clever thoughts. Geraldine Fitzgerald and I worked together on the first series. It was lovely to have someone to share it all with. Michelle Martin, Melissa Hogenboom and Fiona Hill produced the interviews with Sanjeev Gupta, Sandy Knapp and Richard Fortey respectively. Maria Simons holds everything together and somehow manages to remain cheerful, no matter what.

Without Jim Al-Khalii these interviews would not exist. Thank you, Jim for your wonderfully warm interviewing style, for asking questions with humor and humility, and for putting up with me jabbering in your ear. I promise to not drink so much coffee in future. I have enjoyed all the days we have spent in *Nan Winton,* our cubicle on the second floor of New Broadcasting House, wondering how to make the next conversation with a scientist the best one ever. It's been a great privilege to work with you all these years, and I appreciate your patience and, in particular, your kindness.

My brother Will, sister Camilla and mum, Celia have been unfailing encouraging. Eliza and Rosie Quint transcribed many of the interviews and were much better at it than a computer [en dash] thank you for all your kind words of support and for letting me squat in your bedrooms to write. The biggest thank you of all, however, must go to Mike. I am grateful for your intelligent insights and I cherish all your love and support. You have made it possible for me to write this book.

[illegible]

[illegible]

[illegible]

PRAISE FOR *SPIN*

'Anyone with a passing interest in today's politics will raise a wry smile at this Ben Elton-like comedy of spin by someone very much in the know'
Daily Mail

'A searingly accurate and far from encouraging account of life in government under new Labour. But *Spin* is also fast-paced, entertaining and a good yarn'
Ann Widdecombe, *New Statesman*

'Highly readable, intriguing . . . there is a bewitching, persuasive darkness to the British political class in *Spin* . . . How frightening that this "novel" is so believable'
Mail on Sunday

'The leaking, counter-leaking and blackmail are terrific . . . Sixsmith is a master of storytelling'
Daily Telegraph

'The book heaves with details that will fascinate'
Guardian

SPIN

Martin Sixsmith was born in Cheshire and educated at Oxford, Harvard and the Sorbonne. From 1980 to 1997, he worked for the BBC, where he was the Corporation's correspondent in Moscow, Washington, Brussels and Warsaw. From 1997 to 2002 he worked for the government as director of communications and press secretary first to Harriet Harman, then to Alistair Darling and finally to Stephen Byers. He is now a writer, presenter and journalist.